THE SOCIAL AUDIT POLLUTION HANDBOOK

Other Social Audit Publications
SOCIAL AUDIT 1
SOCIAL AUDIT 2
SOCIAL AUDIT 3: TUBE INVESTMENTS LTD
SOCIAL AUDIT 4
THE ALKALI INSPECTORATE
SOCIAL AUDIT 5: CABLE AND WIRELESS LTD
SOCIAL AUDIT 6: COALITE AND CHEMICAL PRODUCTS LTD
SOCIAL AUDIT 7 and 8: AVON RUBBER CO LTD

THE SOCIAL AUDIT CONSUMER HANDBOOK:
Charles Medawar

TD
194
.6

THE SOCIAL AUDIT POLLUTION HANDBOOK

HOW TO ASSESS ENVIRONMENTAL AND WORKPLACE POLLUTION

MAURICE FRANKEL

Distributed in the U.S.A. by
Humanities Press
Atlantic Highlands, N.J.

© Maurice Frankel 1978
All rights reserved. No part of this publication may be reproduced or transmitted, in any form or by any means, without permission

First published 1978 by
THE MACMILLAN PRESS LTD
London and Basingstoke
Associated companies in Delhi
Dublin Hong Kong Johannesburg Lagos
Melbourne New York Singapore Tokyo

Typeset, printed and bound
in Great Britain by
REDWOOD BURN LIMITED
Trowbridge & Esher

British Library Cataloguing in Publication Data

Frankel, Maurice
 The Social Audit pollution handbook
 1. Environmental impact analysis—Amateurs'
 manuals 2. Pollution—Great Britain—Measurement—
 Amateurs' manuals
 I. Title
 614.7   TD194.6
 ISBN 0-333-21646-6
     0-333-21647-4 Pbk

The paperback edition of this book is sold subject to the condition that it shall not, by way of trade or otherwise, be lent, resold, hired out, or otherwise circulated without the publisher's prior consent, in any form of binding or cover other than that in which it is published and without a similar condition including this condition being imposed on the subsequent purchaser.

This book is sold subject
to the standard conditions
of the Net Book Agreement

Social Audit

Social Audit Ltd. is an independent, non-profitmaking organisation concerned with improving government and corporate responsiveness to the public generally. Its concern applies to all corporations and to any government, whatever its politics.

Social Audit Ltd. is also the publishing arm of **Public Interest Research Centre Ltd.**, a registered charity which conducts research into government and corporate activities.

These two organisations are funded mainly by grants and donations and through the sale of publications. Their work has been supported mainly by the Joseph Rowntree Social Service and Charitable Trusts. In addition, support for individual projects has been received from The Social Science Research Council, The Ford Foundation, Consumers' Association, The Allen Lane Foundation and other individuals and institutions.

The name **Social Audit** derives by analogy: if there are financial audits—regular reports on the way a company performs its duty to shareholders—why should there not be 'social audits'—reports for employees, consumers, indeed for everyone affected by what a company does?

But **Social Audit** is not only concerned about companies; it believes that democracy is debased by lack of accountability in government and in any other major centre of power. Social Audit argues that people must be allowed to know about the decision-making done in their name—and must then be allowed to play the utmost part in controlling their own lives.

Between 1973 and 1976, our reports were published in the journal **Social Audit** which was made available only to subscribers. Reports are now published on a one-off basis and made generally available.

Each issue of the journal **Social Audit** contains three reports—or their equivalent—and each report runs from about 5000-10 000 words in length.

If you would like to place an order or receive further information, please write to Social Audit Ltd., 9 Poland Street, London W1V 3DG, England

Acknowledgements

Public Interest Research Centre gratefully acknowledges the support received from the Ford Foundation towards the preparation of this Handbook.

We are also grateful to the Joseph Rowntree Social Service and Charitable Trusts for their continuing assistance and to the Elmgrant Trust and other individual supporters.

I would particularly like to thank Angela Kaye and Andrew Wade for their help with research. I am grateful to all those who read and commented so helpfully on the manuscript: Roger Breach, Alan Dalton, Charles Medawar, Sam Radcliffe, David Rennie, John Rhoades, Roger Tagg, Edward Ward and Peter Williams. Their advice is greatly appreciated, though of course they are not responsible for opinions expressed in the Handbook, or for any errors or omissions.

We and Macmillan also wish to thank the following who have kindly given permission for the use of copyright material: The Controller of Her Majesty's Stationery Office for the extracts from HMSO publications; C.R.C. Press Inc. for a table from the *Handbook of Environmental Control, Vol. 4, Wastewater Treatment*, by R. C. Bond and C. B. Straub and William Heinemann Medical Books Ltd for a figure from *Enviromental and Industrial Health Hazards, A Practical Guide*, by R. A. Trevethick. The Institution of Water Engineers and Scientists for a table from the Paper presented by D. J. Brewin, M. S. T. Chang, K. S. Porter and A. E. Warn at the Final Proceedings of the *Symposium on Advanced Techniques in River Basin Management: The Trent Model Research Programme*, Birmingham University 1972; and a table from the Paper presented by D. G. Miller of the *Symposium on Development in Water and Sewage Treatment*, 1976. The National Water Council for a table from the *Review of Discharge Consent Conditions*, February 1977, slightly amended by the National Water Council, 1978; the North West Water Authority for the extract from their *Water Quality Review 1975* and the Water Research Centre for a table from *Notes on Water Pollution No. 65*.

Every effort has been made to trace all the copyright holders but if any have been inadvertently overlooked the publishers will be pleased to make the necessary arrangement at the first opportunity.

Contents

Social Audit   v

Acknowledgements   vi

Introduction   ix

Toxic Hazards

1   Health Hazards and Standards   3

2   Investigating Hazards   18

Air Pollution

3   Air Pollution Law and Standards   61

4   Air Quality Objectives   84

5   Air Pollution Monitoring   95

Water Pollution

6   River Pollution Law and Standards   117

7   Water Quality Objectives   132

8   River Pollution Monitoring   149

9   The Pollution Audit   161

Appendix 1   Published Sources of Information on The Composition of Trade-named Products   169

Appendix 2	Ordering US Government Publications	172
Appendix 3	Water Authorities and River Purification Boards	173
Appendix 4	Conversion Factors	175
Appendix 5	Social Audit: Contents of Reports 1973–76	177

References and Notes 181

Index 198

Introduction

This Handbook has been designed as a layperson's guide to information about toxic hazards occurring:

inside factories
in the outside air
in rivers and in drinking water

The Handbook has been written, in particular, for workers and their representatives, for people living near polluting factories, and for anyone concerned with the social and environmental impact of industry. It attempts to show readers who have no special scientific knowledge how to uncover, interpret and use information about the toxic hazards found in or around industry.

The Handbook explains how to find out:

(1) What substances are used in or discharged from a particular factory.

(2) What concentrations of toxic substances exist inside, or in the air and water around, the factory.

(3) Whether the concentrations of pollution inside or around the factory are likely to harm human health or damage the environment.

(4) Whether factories are complying with legal standards regulating the discharge of pollution.

(5) How much protection official 'safe limits' really give.

The Handbook does not refer specifically to problems caused by the disposal of waste on land, explosive hazards, nuclear hazards or marine pollution—but many of the principles described, and in particular the chapter on investigating hazards, will be applicable to these topics. The Handbook's use is not limited to Britain, although its account of the law and some of the sources of information are specific to the UK.

Information and secrecy

Recent laws have acknowledged the right of the public to obtain information about the hazards they face at work or in the environment. Nevertheless, although the wisdom of public and worker 'participation' is widely acclaimed, managers in industry and government officials are sometimes reluctant to accept that the layperson could ever understand—or even have a legitimate reason for requesting—information about actual hazards. This view was expressed by the head of the government air pollution agency, the Chief Alkali Inspector, in 1968. 'Abating air pollution is a technical problem' he wrote in a memorable phrase, 'a matter for scientists and engineers, operating in an atmosphere of co-operative officialdom'.[1]

Similar views are sometimes found in industry. To take just one example: in 1977, we asked several suppliers of chemicals used in industry how they would respond to requests for information about the hazards of their products received from workers using them in industry. The Health and Safety at Work Act requires suppliers to inform their customers about toxic hazards of their products, but does not oblige them to provide information to anyone else. The chief chemist of one firm, the Chemical Division of Uniroyal Ltd., replied:

> The problem with trade unionists is that they are fairly simple souls who've got their information secondhand. This means that a scare can be caused fairly easily. A letter [in reply] would be a dangerous thing—I think we'd have to try and see them personally . . . I wouldn't give them any information in writing because of the ambiguity it could cause. And I would check with the customer first . . . I would want to know why they are asking the question . . . Our customers normally have technical people [who deal with these questions] . . . Then we would probably go and see them . . . we would tend to try and see the officials of the trade union first because those people tend to be a little less militant.[2]

Some companies tend to be more willing to provide information about their activities, following the advice of a former president of the Society of Chemical Industry to 'do right and tell people about it'.[3] Unfortunately, some may stop telling people about it when they do wrong, or at least when they do nothing.

An unusual example of corporate openness was provided by the Avon Rubber Company which, in 1974, agreed to become the subject of an independent report by Social Audit on the impact of its activities on employees, consumers and the environment.[4] The company provided the researchers with a good deal of normally confidential information and asked government departments to open their files on the company to Social Audit. Two examples from this study illustrate firstly how important it is for members of the public or the workforce to be able to independently verify company statements and, secondly, how even an 'open' company instinctively withholds information likely to embarrass it:

> Trade unions representing Avon employees had been told that the company's policy was to avoid using carcinogens—cancer-causing substances—in its factories. The company told its unions that it could give 'a categorical assurance that no raw materials are used anywhere in the Avon Group which might have *any* risk of cancer'. An examination of the company's list of raw materials revealed that it contained 19 suspected or confirmed carcinogens.

> Avon Medicals provided researchers with the results of three analyses of its trade effluent discharges, made by the water authority in 1974. The Chemical Oxygen Demand (COD) of these samples ranged from zero to five parts per million—compared to a legal limit of 600 ppm. Only when Social Audit contacted the water authority directly did it learn that a fourth test had been carried out during this period. The results of this test—not supplied by the company—showed that the COD of the effluent had been 5735 ppm, nearly ten times more than the permitted limit.

Members of the public or workers trying to learn about the toxic hazards they face from industry will always find themselves frustrated so long as those who take decisions can protect themselves from informed comment or criticism by secrecy. 'Knowledge or lack of it is a powerful management tool', a British Leyland executive told a conference in 1975. 'Information is an important determinant of status . . . Management must manage and authority must be maintained'.[5]

The manager's monopoly of information on toxic hazards has been shaken by two recent pieces of legislation. The 1974 Health and Safety at Work Act requires employers and factory inspectors to keep workers informed about the hazards they face at work; the 1974 Control of Pollution Act requires water authorities to publish details of firms' water pollution discharges and allows local authorities to publish information about air pollution emissions.

This Handbook is designed to help workers and other members of the public find and understand information about hazards from industry; and having understood it, be able to appreciate, question, and if necessary protest at the actions of those responsible for controlling hazards on our behalf.

Toxic Hazards

This section of the Handbook deals with toxic hazards in the workplace and environment. It is divided into two chapters.

Health Hazards and Standards (chapter 1) explains how the toxic hazards of substances are studied and how the results of surveys and experiments are used to set limits for human exposure. It explains the limitations of the research methods and of standards based on them and suggests how you can evaluate the degree of protection given by 'safe' limits and exposure standards.

Investigating Hazards (chapter 2) describes the main sources of information on the hazards of toxic substances and tells you how to make full use of your legal rights to be informed. The chapter explains how to identify a substance in the scientific literature and to discover the chemical ingredients of a substance known only by its trade name. It describes the main reference books on toxic hazards, illustrates their usefulness with extracts from each, and tells you how to bring the information contained in them up to date. With this chapter you should be able to discover all that is known about the hazards of a substance used at work or found in the environment.

1 Health Hazards and Standards

The factory air quality standard for cyclohexanone, a common industrial solvent, has been set on the basis of only two reports of its effects. Both were published in 1943; only one involved observations on human beings.

The first report found that concentrations of 190 parts per million of cyclohexanone in the air 'induced just demonstrable degenerative changes in the liver and kidneys of rabbits'. The report that dealt with human beings found that 'cyclohexanone was not tolerated at 50 parts per million, throat irritation being the most marked effect'.

On this evidence, the Threshold Limit Value for cyclohexanone, the level to which workers may be continuously exposed throughout their working lives, was set at 50 parts per million.[1]

Summary

The hazards of a toxic substance to man can only be reliably demonstrated after it has come into use and has had an opportunity to injure people exposed to it. Methods of predicting hazards without waiting for human casualties depend on animal experiments—but the results of these are always difficult to apply precisely to human exposure.

No experiment can ever prove that a given concentration of a substance is 'safe'—it can only show that certain ill-effects seem to be absent. As more research is done, and methods of investigation improve, many 'harmless' substances or concentrations are found to be hazardous.

Pollution control inside and outside factories attempts not to eliminate toxic substances but to keep them to low concentrations which are thought to be harmless. However, exposure standards do not usually try to give *absolute* protection. They take the costs of controlling pollution into account when setting limits. The standards may allow some risk to health if the cost to industry of avoiding it is thought to be too great.

The importance of concentration

Almost any substance can be harmful to human beings if it enters the body in large enough quantities. Substances that are normally thought of as 'safe'—for example, aspirin—can be lethal in excess. Similarly, minute concentrations of even the deadliest poisons may be relatively harmless: traces of arsenic are present naturally in sea water.

The aim of pollution control—both outside and inside the workplace—is not normally to eliminate toxic substances but to keep them within 'safe' levels.

The potential hazard of a substance depends on the likelihood that a damaging dose of it can reach a vulnerable target in the body. Many substances only attack certain organs or tissues. Their toxicity depends on the route by which they enter the body: a dose that would damage the lungs if inhaled may be harmless if swallowed.

To some extent, the body can resist attacks by toxic substances. The mucus and fine hairs in the respiratory system help to keep dusts from reaching the lungs—though the defences can be damaged by smoking; the liver and kidneys help to break down and eliminate harmful substances that reach the blood. These defences may be weaker in some individuals than in others, either because they are generally less healthy or because they are born with less effective natural defences. Thus, concentrations of pollution that can be tolerated by most of the population may be lethal to a certain proportion of people, especially to the sick, the very old or the young (children are known to be as much as five times more sensitive than adults to certain compounds).[2] The notorious London smog of 1952 led to nearly 4000 deaths, brought on by the air pollution in people who, on the whole, were already suffering from respiratory or heart disease.

A single toxic substance may have both immediate and long-term effects. The immediate (*acute*) effects are usually caused by a single exposure to a relatively high concentration of the substance: an acid splash burning the skin, or dizziness and nausea after breathing benzene vapours. However, continued exposure to perhaps lower concentrations may produce completely different (*chronic*) effects which are much less obviously related to the cause. Traces of acid on the skin may give rise to skin disease; breathing benzene fumes in over several years may lead to leukaemia.

A substance may show new, or vastly increased, toxic properties in the presence of other substances. Certain combinations of pollutants are known to be much more harmful when they are both present together than their individual toxicities would suggest. This is known as *synergism*, and has been found to occur for mixtures of smoke and sulphur dioxide in the air (see page 86), chromium and nickel in water (see page 138) and for other combinations.

To understand the hazards of a toxic substance we need to know, firstly, what kinds of damage it can lead to, both in the short-term and in the long-term, and, secondly, how the damage is related to the concentration of the substance and the length of exposure.

Is there a no-effect level?

Before a standard for human exposure to a toxic substance can be set, we ideally need to know the complete 'dose–response' relationship; that is, the effects caused by exposure to the whole range of concentrations normally found. As the concentration decreases, those harmful effects being studied may seem to disappear: *but no scientific test is capable of proving beyond any doubt that a particular concentration of a substance causes no ill-effects at all.*[3]

As methods of studying hazards improve, substances previously thought to be harmless are found to cause injury, and harmful effects are detected at concentrations once thought to be safe.

In 1958, Professor Lawther of the Medical Research Council in London, reported that the condition of bronchitis patients had been found to worsen when the daily concentrations of smoke and sulphur dioxide in the air simultaneously rose above 300 and 600 micrograms per cubic metre ($\mu g/m^3$) respectively.[4] These were the lowest levels at which ill-effects had been detected at that time. After continuing and refining their techniques, Lawther and his colleagues reported, in 1970, that they had detected a worsening in the condition of bronchitis sufferers when daily smoke and sulphur dioxide concentrations exceeded 250 and 500 $\mu g/m^3$.[5]

Experiments may sometimes overlook harmful effects because they are based on too small samples of animals or people. Suppose that a certain dose of a chemical caused cancer in 0.1 per cent of people or animals exposed to it. Cancer could be expected in one out of every 1000 animals used in laboratory tests: experiments using less than several thousand animals could easily fail to produce even a single case of cancer. To call the concentration tested in this way 'safe' or the 'no-effect level' would, according to the US National Academy of Sciences (NAS) be 'statistically meaningless . . . [it] is completely compatible with the presence of an adverse effect, which in further studies with larger sample sizes or with different types of observation might lead to a positive outcome'.[3] Instead, the NAS recommends that the term 'no observed effect' should be used, provided it is qualified by a statement explaining the size of the group in which no effect was found.

For some substances there may be what is known as a 'biological threshold', a concentration below which the human body is able to break down or eliminate the whole dose reaching it without suffering any harmful effects. Above this threshold some harm occurs even if it is undetectable by normal methods.

However, there are no known 'thresholds' for substances that cause cancer—carcinogens. If the concentration of a carcinogen is decreased, the number of cases of cancer found drops—but as long as the substance is present at all, it is possible that some individuals who would not otherwise be affected will develop cancer.

Testing toxic substances

Standards that set permitted concentrations for human exposure to toxic substances are based on information from three kinds of sources: (a) studies observing small numbers of people—either human volunteers or workers in industry; (b) surveys of large numbers of people known to have been exposed to the substance in their normal lives; and (c) laboratory experiments with animals.

In each case, the health of the group exposed to the substance is compared to the health of an identical non-exposed group. Statistical tests are used to decide how likely it is that any difference in the response of the two groups is caused by the substance under study and not by chance.

It is useful to understand the limitations of these test methods before trying to use any standard for human exposure.

Tests on human volunteers

The most useful information about the effects of toxic substances on man comes from experiments in which human volunteers are exposed to known concentrations of pollutants.

Since researchers cannot expose volunteers to dangerous concentrations of chemicals, there are relatively few studies of this sort, and those that have been done are of limited scope. Substances can only be tested in concentrations that are known to produce minor and reversible effects in animals. Volunteers have to be fit adults in good health, and are therefore unrepresentative of the general population. Even so, they still run the risk of serious injury:

> After an experiment in 1957 which exposed healthy adult men to sulphur dioxide and sulphuric acid mist, one of the volunteers developed a leakage of fluid from an inflamed lung.[6] Both of the researchers who organised the tests became highly sensitive to the pollutants involved and one 'developed a moderately severe but extremely persistent bronchitis which was immediately exacerbated into an uncomfortable period of coughing and wheezing on exposure to either sulphur dioxide or sulphuric acid'.

A more common source of human test data comes from workers who have been exposed to previously untested chemicals used in industry.

Around 40 per cent of factory air quality standards (Threshold Limit Values) are based on such observations.[7]

Workers have often gone on providing evidence about the hazards of chemicals long after the lethal working practices they were using should have been stopped. The following extract is from the official documentation of the Threshold Limit Value for benzene:

> Winslow (1927) reported blood changes in workers ... Greenburg (1939) described nine cases (of benzene poisoning) with one death ... Bowditch and Elkins (1939) ... (reported) eleven fatal cases ... Heimann and Ford (1940) found one death ... Wilson (1942) reported three fatal cases ... Elkins (1959) stated that more than 140 fatal cases of benzene poisoning had been recorded prior to 1959. Vigliani and Saita (1964) listed 26 deaths from chronic benzene poisoning in two provinces in Italy between 1960 and 1963.[1]

Surveys of exposed people

Another way of discovering how toxic substances affect man (or other species) is to examine the health of large numbers of people who have been exposed to known concentrations: for example, the whole population of a town suffering from heavy air pollution, or all workers in a particular industry. This technique is known as *epidemiology*, from the same root as 'epidemic', and means the study of disease in the population.

The advantages of this method are that it looks at the real world instead of the laboratory; it deals in large numbers of people and not with handfuls of possibly unrepresentative individuals, and it involves human beings and not experimental animals.

Its great and overwhelming drawback is that it can only be used *after* a toxic substance has come into use and has caused hundreds or thousands of injuries or deaths. It does not allow the effects of new substances to be predicted *before* they are released into the workplace or environment.

Even 10 to 20 years after a substance has first come into use, when the long-term ill-effects are beginning to appear in exposed people, surveys may be made impossible because no-one kept accurate records in earlier years. Many cases of industrial disease have been—and probably still are—misdiagnosed or unreported. Pollution levels in the past may not have been monitored, or if they were, records were not kept. One result of this has been that health surveys of people exposed to air pollution in the UK deal mainly with smoke and sulphur dioxide hazards—though far more toxic pollutants are common—because these are the only two substances that have been monitored nationally.

Animal experiments

The only way of estimating how dangerous a new substance may be—without waiting for the first human casualties—is to test it on animals.

In the past, it has been quite normal for new chemicals to be introduced into industry, or discharged into the environment, without any form of toxicity testing. The Health and Safety at Work Act now requires manufacturers to test chemicals before they are supplied,[8] and the Health and Safety Executive has suggested a statutory testing scheme for new chemicals—though it will not apply to the thousands of untested chemicals already in use.[9]

TABLE 1.1
Time necessary to produce death in animals exposed to high concentrations of nitrogen dioxide. Typical results from a single animal experiment[10]

Concentration, mg/m^3	Time until death min	% deaths
56·4	—	0
188·0	318	74
282·0	90	70
752.0	58	92
1128·0	32	93
1524·0	29	100
1880·0	19	100

TABLE 1.2
Response of different animals to short-term exposure to nitrogen dioxide gas[10]

Species	Concentration (parts per million)	Exposure (hours)	Effects
Rats	1·0	1	Some changes in certain lung cells. Reversible
Mice	5·0	2	Increased frequency of death after infection with pneumonia-causing bacteria
Mice	7·7	6	20 per cent decrease in voluntary running activity
Monkeys	15·0	2	Decrease in volume of air taken in with each breath; changes in cells of liver and kidneys
Monkeys	35–50	2	Marked increase in breathing rate and decrease in volume breathed in. Collapse of air sacs in lung; inflammation of air passages. Development of fibrous tissue in heart and changes in kidney and liver cells
Various	100	5	74 per cent of test animals killed

Animal experiments can ideally help to build up a full dose–response relationship for a toxic substance. In practice, experiments often look at only two or three different doses of the chemical, and instead of examining all the possible toxic effects (such as changes in behaviour, growth, reproduction, body chemistry and tissue structure) will look at a single key response—often the death of the animal in a given time (see Table 1.1). When all the different results obtained from different experiments are put together, the *beginnings* of a dose–response relationship may be revealed (see Table 1.2).

Applying animal results to man

Tests on mice and rats are all very well if our ultimate aim is to set rodent-welfare standards. However, most research with animals is done to set standards that will protect *human* life; and there are many difficulties in relating animals results to human beings.

The species of animal tested may respond quite differently from human beings. Some animals show the same response as man to certain chemicals—others do not. For example, thalidomide, the drug that led to the birth of deformed children after it had been taken by pregnant women, was tested on pregnant rats—who later gave birth to undeformed, though fewer than normal, offspring. If it had been tested on rabbits—one of the few species that reacts like human beings to this drug—researchers would have been warned beyond any doubt that this drug was too dangerous to release.[11] Anticipating which species of animal is likely to react like man to a particular substance is a common problem:

> when niobium ion was tested in the mouse it was found to decrease the growth rate but increase the longevity of the animal. However, when tested in the rat, the growth rate was significantly increased while the longevity was decreased. In neither species was survival affected. In trying to extrapolate to man, there is no basis for determining whether man reacts like the rat or mouse.[12]

The experiment may be designed for the convenience of researchers rather than for its relevance to man. Most animal experiments test very high doses of substances and only look for the short-term effects. Often the effect chosen is the death of the animals although human beings are rarely exposed to concentrations of pollution that will kill them in a few days: they are much more likely to be faced with lower levels that will either produce less drastic short-term effects or serious illness after long-term continued exposure.

Industry may be reluctant to fund longer-term toxicity testing of new chemicals, because it is expensive. A relatively modest study, in which animals are exposed to only three different concentrations of a substance in the air for 90 days costs £20 000 (at 1976 prices); a two-week experiment to find the lethal dose of a chemical can be done for around £500.[9]

A very rough method of predicting from animal experiments how humans will react to toxic substances has sometimes been used. The method depends on a number of assumptions that may not always be valid. It assumes that human beings will show the same kinds of ill-effects as animals after exposure to a toxic substance, and that the human responses will be found at proportionally higher or lower doses. If we know the concentration at which each different effect occurs in the animal, and the concentration at which just one of these effects is found in man, then the point at which all the other responses will be found in man can, if the assumptions are true, be predicted.

This procedure has been used to set some human exposure standards:

> Hospitalized volunteers with terminal disease not involving those organs and tissues known to be affected by the test substances may be exposed at or around the level at which the most sensitive response test in animals was positive. Such a procedure was used prior to the development of the air limit for uranium. It was found that man responded at a level 10-fold higher than that of the most sensitive species (the rabbit) as judged by the most sensitive test of uranium toxicity . . . Thus a known minimal safety factor can be established.[7]

Animal tests used to detect cancer-causing substances can be applied more directly to human beings. These experiments are not primarily concerned with the concentration of a substance that causes cancer—there is thought to be no 'safe' level for a carcinogen—but with discovering whether or not the substance induces cancer at all. Chemicals are tested on animals in massive doses, far greater than those that people are likely to encounter normally, to help magnify the effects of a 'weak' carcinogen that would otherwise be undetected unless hundreds or thousands of animals were used. If animals develop cancer, the substance is treated as a probable human carcinogen.[13]

A new, quick method of detecting possible carcinogens has recently been developed based on the observation that a chemical that causes mutations in bacteria is very likely to also produce cancer in animals and presumably in man. The attraction of this method is that 'mutagenicity' tests with bacteria can be carried out quickly and inexpensively: a chemical can be tested in just three days at a cost of perhaps £500, whereas a full-scale test using laboratory animals lasts for the lifetime of the animals and may cost £100 000. Although the new method is not infallible, it will for the first time make it possible to screen large numbers of chemicals for carcinogenicity.[14]

An example of a survey result

A typical example of the kind of evidence produced by research into toxic hazards is shown in the following statement. It comes from a study which examined the health of children living in areas polluted by nitrogen dioxide (NO_2) gas in Chattanooga, Tennessee: '[*A*] *greater frequency of acute bronchitis was observed when the mean 24-hour NO_2 concentration, measured over a 6-month period, was 118 µg/m^3 (micrograms per cubic metre)*'.[10]

To understand what this finding means, you have to remember how studies of this sort are carried out. In this case, the frequency of certain kinds of illness in a high-NO_2 area was compared with that in a 'control' area with a low concentration of NO_2. Reading the report of the study shows that air in the control area contained 81 µg/m^3 of NO_2.

The study shows that certain kinds of illness are linked with, and presumably caused by, continued exposure to air containing 118 µg/m^3 of NO_2.

The study does not show that concentrations of NO_2 below 118 µg/m^3 are harmless. It did not look at areas with lower NO_2 levels. If it had done, it might have found that there was also an increased frequency of bronchitis in areas with, say, 100 or 90 µg/m^3 of NO_2.

The study does not show that NO_2 concentrations found in the control area (81 µg/m^3) are harmless. They are less likely to cause bronchitis than the higher concentrations studied but they may nevertheless cause more bronchitis than still lower concentrations, or no nitrogen dioxide at all.

Even if there had been no increased frequency of bronchitis in areas with 118 µg/m^3 of NO_2, this concentration could still not be assumed to be 'safe'. The pollution could be causing an increase in any one of the many different kinds of diseases or symptoms not examined in the study. Or it may have been causing an increase in bronchitis in a sector of the population not covered by the study, for example in elderly people. To save having to carry out separate studies on all sectors of the population and looking for all possible symptoms, researchers will usually try to study a group known to be particularly vulnerable (as, for example, were the young children covered in this study) and look for the most easily produced harmful symptoms—perhaps those suggested by animal experiments.

Standards for human exposure

The hazards of many common industrial chemicals and environmental pollutants have only been partially—and inadequately—studied. It is nearly always impossible to say at what concentration and period of exposure the very first signs of ill-effects in man just begin to appear.

This makes it difficult to set precise safety standards. Standard-setting bodies in the UK repeatedly stress the importance of avoiding unnecessary costs when setting exposure limits. The Department of the Environment has warned that, in the absence of proper information about hazards, 'fear of the unknown may exercise undue influence on decisions about standards'. It explains that the British system aims to 'ensure that the adverse effect on man is the lowest practicable, having regard to the costs involved'.[15]

Standards used in the UK to protect people from hazards caused by air pollution, river pollution or drinking water are described in chapters 4 and 7. This section describes the standards used to limit exposure to toxic substances in the workplace. It illustrates the two basic qualifications that are attached to nearly all exposure standards, namely that **health protection standards (a) do not give, or claim to give, complete protection to health, and (b) the degree of protection is a political decision balancing the costs of prevention against the benefits.**

Threshold Limit Values

Many countries in western Europe, including the UK, base their standards for workplace pollution on the Threshold Limit Values (TLVs) set for use in the USA by the American Conference of Governmental Industrial Hygienists (ACGIH). The ACGIH, despite its title, is not a government body, but the professional association of industrial hygienists, many of whom work in industry.

With only a few exceptions, the American TLVs are used in the UK, and the Health and Safety Executive reprints the TLVs in one of its guidance notes.[16] They are not themselves legally enforceable, although the Factory Inspectorate uses them as a guide.

The TLV is the concentration of workplace pollution to which 'nearly all workers' may be repeatedly exposed for eight hours a day during a 40-hour week without, in the view of the ACGIH, suffering harmful effects. For most substances, the TLV is not a maximum that may never be exceeded, but an average. It allows the concentration in the factory to be above the limit for a part of the day providing that the high levels are compensated for by equivalent low levels at other times. However, Short Term Exposure Limits—setting a maximum concentration that should not be exceeded for more than 15 minutes continuously—were introduced by the ACGIH in 1976 and some substances have 'ceiling' values, which should not be exceeded *at all* even instantaneously.

Note that the ACGIH does not claim that the TLVs will protect everyone:

> a small percentage of workers may experience discomfort from some substances at concentrations at or below the threshold limit; a smaller percentage may be affected more seriously by aggravation of a pre-existing condition *or by development of an occupational illness*[16] (emphasis added).

These standards therefore allow some hazards to workers—and at times quite serious hazards—if the costs of avoiding factory pollution are thought to be too great.

By contrast, standards set in the USSR are based solely on scientific evidence of harm and do not take into account the costs or technical difficulties of achieving them. The Soviet standards aim to avoid causing 'any diseases or deviations from a normal state of health detectable by current methods of investigation'.[17] In some cases these standards are set partly on the basis of evidence of the concentrations at which the body detects foreign substances, even though its response may be reversible and apparently cause no ill-effects.

TABLE 1.3
Factory air standards used in the UK, Sweden and the USSR[18]

	UK TLV (1974)	Sweden (1975)	USSR (1972)
	(all figures in mg/m^3)		
Acetaldehyde	360[a]	90	5
Acetone	2400	1200	200
2-Butanone	590	440	200
Carbon monoxide	55	40	20
Ethylchloride	2600	—	50
Manganese and compounds (as Mn)	5(C)[b]	2·5	0·3
Propylene oxide	240	—	1
Toluene	750[c]	375	50
Trichloroethylene	535	160	10

Notes

[a] This limit had been reduced to 180 mg/m^3 by 1976.[16]

[b] 'C' indicates that the limit is a 'ceiling' limit that may not be exceeded even instantaneously. All USSR standards are 'ceiling' values—most TLVs are average limits that may be exceeded for a time providing the concentration is below the limit by an equivalent amount for the same length of time during the day.

[c] This limit had been reduced to 375 mg/m^3 by 1976.[16]

Many of the standards set in the USSR are therefore considerably stricter than the TLVs used in the UK. For about a third of the substances for which standards existed in both countries in 1974, the UK limits allowed over ten times more of the toxic substance than the Soviet standards.[18] Eighteen of the Soviet standards allowed only 1/50th of the amount permitted under the TLV, while in one case the Soviet standard was 1/240th of the TLV[19] (see Table 1.3). The Soviet standards are even more rigorous than these figures suggest since, unlike the TLVs, they are maximum limits that may not be exceeded for even the shortest time.

These standards are probably not strictly enforced in the USSR. Reports are often published describing ill-health in workers exposed to concentrations of factory pollution that exceed the Soviet standards, and Soviet writers have acknowledged that some of their air pollution standards, at least, are not technically attainable at the present time.[20]

However, they do help to illustrate how far the TLVs used in this country are from being 'safe' standards.

The evidence for TLVs

The evidence on which each TLV has been set has been published in a book called *Documentation of the Threshold Limit Values*.[1] Supplements to this book are published from time to time because TLVs are changed periodically—and usually made stricter—as new evidence is reported.

The *Documentation* shows that although quite a lot is known about the hazards of some of the 500 or so substances for which TLVs have been set, there is a startling lack of evidence for others.

In some cases the TLV is based largely—or wholly—on the results of animal experiments, with no real evidence about human exposure.

The TLV for chlorine trifluoride is based solely on the results of experiments with dogs and rats. Two dogs, exposed to 1·17 ppm (parts per million) of the substance for six months, both developed pneumonia and one died. Six out of 20 rats exposed to the same concentration also died. A ceiling TLV of 0·1 ppm was set.

The TLV may be set at a level at which no tests or observations have been reported. In these cases, harmful effects have usually been detected at certain levels and a lower 'safe' level is deduced—largely by guessing.

Chloropicrin is a powerful irritant, developed as a 'vomiting gas' for military warfare. It is used as a pesticide or in industry as a warning agent. The TLV is based on warfare studies published before 1931. Four parts per million of chloropicrin 'renders a man unfit for activity' while 0·3 ppm produced 'painful irritation to the eyes in 3 to 30 seconds'. No reports suggesting that lower concentrations could be tolerated are mentioned. The TLV was set at 0·1 ppm.

In some cases the TLV is set at a concentration which is known to cause irritation or discomfort to humans.

2-Butanone, also known as methyl ethyl ketone, is an industrial solvent. The TLV for this substance is based on four reports, the most recent published in 1950. These studies found slight nose and throat irritation occurred in people exposed to 100 ppm, mild eye irritation in some subjects at 200 ppm, and 'low grade intoxication' at 300 to 600 ppm. A report published by the manufacturers of this substance stated that 'the highest concentration . . . which the majority of human subjects estimated satisfactory for eight hours was 200 ppm'. The TLV was set at 200 ppm.

Sometimes the TLV has been set at a concentration which has been reported to produce serious injury or even death in exposed workers.

Chromic acid and chromate compounds are extensively used in industry, especially in metal plating processes. The TLV is based on 11 reports, a relatively large number.

More than half the workers covered in a 1953 study of the US chromate industry were found to have wounds in the nasal septum (the cartilage between the nostrils) after exposure to varying concentrations of chromates, which in some areas were below 0·1 mg/m^3. After exposure to between 0·11 and 0·15 mg/m^3 other workers were found to suffer from ulcers of the nasal septum, irritation of the mucous membranes and chronic asthmatic bronchitis. One case of cancer of the nasal septum and one case of lung cancer was later found amongst these workers. In another study, seven workers developed lung cancer after exposure to between 0·01 and 0·15 mg/m^3 of chromates.

The TLV was set at 0·1 mg/m^3. The ACGIH nevertheless believes that this limit has been able to prevent new cases of damage to the septum and lung cancer.

Not all TLVs are as suspect as these—and some are backed by very much more convincing evidence. These examples have been chosen because they illustrate how inadequate standards can become under a system which, according to the chairman of the TLV committee, aims to 'provide safety without overprotecting the worker . . . which private enterprise can afford only at the expense of the consumer public which pays the increased cost of their product'.[7]

Applying health standards

No standard for exposure to toxic substances—other than complete elimination—can be guaranteed to be completely safe. Health protection standards often sacrifice some possible protection for a certain saving in costs. In any case, the evidence on which standards are based is usually inadequate: you can be certain that many substances thought to be harmless today will in future be shown to cause—and will now be causing—serious injury to health. The Health and Safety Executive recommends that 'the best working practice is to keep concentrations of all airborne contaminants as low as practicable whether or not they are known to present a hazard, and irrespective of their TLVs'.[16] The key word in this statement is 'practicable'—which means as much as is possible at a reasonable cost. Feel free to challenge other people's definitions of 'practicable'.

Standards for toxic substances should never be taken as a licence to pollute a clean workplace or environment up to the permitted limit. The standards themselves usually say this, but the warning—if it is read at all—is usually forgotten.

TLVs and standards derived from them are in no sense 'acceptable' limits for pollution. The Deputy Director-General of the Health and Safety Executive has stated: 'There's no question of (TLVs) being acceptable—what they represent is the threshold of unacceptability'.[21] If factory pollution is not well below the TLV, it is far too high. Use the Russian factory air standards if you want an idea of health protection limits that are based on health and not the avoidance of 'unnecessary' costs.

Try and check the evidence of any health protection standard you want to use. Chapter 2 tells you where to find evidence about health hazards—though the *Documentation of Threshold Limit Values* will provide most of what you need on TLVs.

Once you have the evidence:

(a) Find out if any ill-effects in animals or in human beings have been reported at concentrations below those chosen for the standard. If they have, the standard is inadequate.

(b) Look at the lowest concentration at which ill-effects have been reported. This will not necessarily be the point at which harm first occurs, so do not assume that concentrations below this level are harmless. On the other hand, concentrations above this level should be unacceptable.

(c) A study that found no harmful effects at a particular concentration does not mean that this level is safe. It only means that the specific ill-effects looked for were not observed in the people studied.

A study that only looked for a possible increase in deaths amongst healthy adults exposed to a particular concentration of substance—and found none—does not show that this concentration is harmless. It may be causing non-lethal illness in the group; it may be fatal, or very harmful, to people who are already unwell or who are very young or old. The study will not be able to suggest that a substance is relatively harmless unless it has searched for the most easily produced symptoms amongst the most vulnerable sectors of the population likely to be exposed to the toxic substance.

(d) Be very suspicious of evidence that comes solely from experiments with animals. Assume that human beings are very much more sensitive than the animals tested.

2 Investigating Hazards

Summary

This chapter is designed as a guide to sources of information on toxic substances used at work or found in the environment. With it—and a good reference library—you should be able to discover all that is known about the hazards of any toxic substance.
The Chapter is divided into nine parts:

Part 1 tells you what to look for in your search (page 18)

Part 2 helps you identify your substance in the scientific literature on hazards (page 20)

Part 3 shows you what to do if you only know the trade name—but not the proper chemical name—of the substance you are investigating (page 22)

Part 4 tells you which organisations may help you by supplying information (page 23)

Part 5 guides you through the reference books on toxic hazards (page 26)

Part 6 shows you how to bring your information up-to-date (page 37)

Part 7 explains how to use air monitoring to decide how dangerous your work environment may be (page 41)

Part 8 tells you where to find information on air pollution hazards (page 43)

Part 9 tells you where to find information on water pollution hazards (page 50)

Part 1—The search

Never assume that a substance is 'safe' just because its hazards are unknown or because it is not mentioned in the books you use.

There are more than four million different known chemicals. We know nothing at all about the toxic hazards of most of these. The most comprehensive list of toxic substances on which some research had been done by 1975 contained information on under 17 000 different substances.

Even the toxic effects of those chemicals that have been studied are usually not fully understood. Most research has looked at the short-term hazards of brief exposure to toxic substances—the effect of continued exposure for many years has often been overlooked. Experiments nearly always study the hazards of individual substances on their own—but the effects of mixtures of several different substances (which are commonly found and may have different effects from those of their ingredients) are almost never studied. Finally, most research is based on experiments with *animals* and the findings are not directly applicable to *human* exposure.

The limitations of information produced by research into toxic hazards are described in chapter 1 of this book. It is advisable to read this chapter before you look up the hazards of a substance.

Always use the most up-to-date sources of information you can find.

The amount of information known about the hazards of toxic substances is increasing very quickly. A chemical that was thought to be safe last year may be a proven killer this year. Reference books on hazards become out of date very quickly: the 1975 edition of the fullest list of toxic chemicals contained 4000 new substances that had not been mentioned in the 1974 edition.

To compare the value of different reference books, this chapter describes the amount of information they provide on a common industrial solvent called *trichloroethane*. Figures 2.1, 2.2 and 2.5 all state that no harmful effects have been found in man at concentrations of less than 500 parts per million (ppm). But the entries in figures 2.3 and 2.9 show that toxic effects occur at only 350 ppm. The reason for this difference is that the first three books use data published before 1973, when a new study showing toxic effects at concentrations previously thought to be harmless was published for the first time.

Part 6 of this chapter explains how to locate the most recent sources of information on toxic substances.

Decide in advance how much information you are looking for.

This guide tells you how to go about finding full details about the hazards of a toxic substance from a variety of different sources. In some circumstances, you may have to check through several of these sources to see whether they contain useful information; this will often be the case when you are dealing with a relatively new substance or one that is not well covered by the main reference books. In many other cases—especially when the hazards of a substance are well known—a single good reference book may give you all the information you need.

Normally, you will want to know:

(1) the immediate (acute) effects caused by a single exposure;
(2) the long-term (chronic) effects of continued exposure;
(3) in all cases, the concentration or dose of a substance that produced any specified effects;
(4) whether a standard has been set or recommended, limiting the amount of substance permitted in the workplace or environment and, if so, whether such a standard can be expected to give adequate protection (see chapter 1).

Part 2—How to make sure you are not confusing your substance with another of similar name

Completely different chemicals often have very similar names. The name of your substance may differ from the name of something else by only a single letter; for example, arsenite and arsenate or sulphide and sulphite compounds, piperazine and piperidine. To make matters worse, the same substance is often known by several different names or *synonyms*. Ethyl acetate, for example, is also called acetic ether, ethyl acetic ester and ethyl ethanoate.

This variety of names can be very confusing when you begin investigating the hazards of a chemical; different books may use different names for the same substance.

Start your search by writing down all the different names of the substance you are investigating. Many of the reference books described in this chapter list synonyms; *Registry of Toxic Effects of Chemical Substances* (see page 28) is by far the best in this way. The chemical dictionaries listed on page 169 will also give you the synonyms of any substance you look up.

Then, if the first name you have for your substance is not listed in one of the sources you are using, you can check to see whether it is listed under one of its other names.

If you come across a chemical with a name similar to the one you are investigating, you can find out whether it refers to the same substance by:

(1) *checking the synonyms of both names*—if the names refer to the same substance then each will be listed as a synonym of the other;

Investigating Hazards 21

(2) *checking the chemical formulae of both*—they will be identical if the names refer to the same substance.

The chemical formula (sometimes also called the molecular formula) is a set of symbols showing the names and numbers of different atoms that make up one molecule of the substance. (For example, C is used as the symbol for carbon, H=hydrogen, Cl=chlorine, and so on.) The chemical formula of any substance will usually be given in any account of a substance's toxic hazards.

If you are investigating the hazards of trichloroeth*ane*—the chemical used as an example in this chapter—you may come across references to a substance called trichloroeth*ene*. Unless you check, you will not be able to tell whether this refers to a completely different chemical—or whether it is a synonym for your substance, or perhaps it is American spelling, or even just a misprint.

Checking the synonyms and the molecular formulae of the two names shows that they refer to totally different substances:

Trichloroethane	**Trichloroethene**
Synonyms	*Synonyms*
Methyl chloroform	Trichloroethylene
Chlorothene	Ethylene trichloride
Ethane, 1,1,1-trichloro-	1,1-dichloro-2-chloroethylene
Formula	*Formula*
$C_2H_3Cl_3$	$C_2H\,Cl_3$
which means that one molecule of the substance is made up of two atoms of carbon, three atoms of hydrogen and three atoms of chlorine.	which means that one molecule of the substance is made up of two atoms of carbon, one atom of hydrogen and three atoms of chlorine.

Unfortunately, it is possible for different substances to share the same chemical formula. This happens when the same atoms combine in more than one way to form similar—but not identical—compounds called *isomers*. Isomers often have different effects from each other on human health, so it is important not to mix them up.

Isomers are normally distinguished from each other by putting a number, or a Greek letter such as alpha or beta, before the name. **Never overlook numbers or letters that come before or as part of a chemical name—they are a vital part of that name.**

Trichloroethane has two isomers. This chapter is looking at the effects of one of these isomers: *1,1,1-trichloroethane* sometimes also known as α (alpha)-trichloroethane. Don't confuse this substance with its isomer, 1,1,2-trichloroethane, also known as β (beta)-trichloroethane. Their hazards are different and they have different TLVs.

Part 3—What to do if you know only the trade name— but not the chemical name— of a substance

Many industrial chemicals are sold under trade names (Flomax, Armogard, Nonox, BHC 318 etc.) which describe the use of product or give it a code number—but do not refer to its chemical composition. Sometimes a trade-named product contains just a single substance; often it is a mixture of several different chemicals, each one having its own toxic properties. Some blended printing inks sold under single trade names may contain as many as 500 different chemicals.

Trade names are usually not listed in the sources of information on toxic hazards (although one source, *Registry of Toxic Effects of Chemical Substances* (see page 28), lists trade-name products that contain only a single substance). Before you can investigate the hazards of a substance identified only by its trade name you must find out what chemicals it contains.

If the trade-named substance is used at your place of work, ask your employer to find out for you what is in it. He is required by law to give you information of this sort (see page 23). If your employer cannot, or will not, provide you with full details, contact the Factory Inspectorate. They will either give you this information, if they have it, or require your employer to obtain it for you from the manufacturer or supplier of the trade-named product.

The only easy way of discovering the composition of a trade-named product is from your employer or the Factory Inspectorate (if the substance is used in your workplace) or directly from the manufacturer of the substance; all other sources of information are incomplete and tedious to use.

If the substance is not used in your place of work you have no legal right to information about its composition and hazards, though some manufacturers may be willing to supply you with this information (see page 24). You will then have to try and identify its composition from the sources listed on pages 169–71. However, there is no guarantee that they will contain the information you need.

The sources of information on trade-named products on pages 169–71 have been drawn from (a) chemical dictionaries, (b) technical reference books, and (c) trade directories and buyers' guides, produced to help companies contact suppliers of raw materials. If the industry in which the trade-named product is used is not included in the list, you can find out whether a trade directory is published by contacting its trade association. Trade associations are listed in *Directory of British Associations*[1] and *Trade Associations and Professional Bodies in the UK*.[2]

Many of the sources listed are American publications. These will only help you identify trade-named products used in British industry if the products are imported from, or exported to, the USA—and if they are marketed under the same trade-name in both countries.

There are other ways of obtaining information about trade-name products. The product may have been described in the trade press when it was first put on the market; if so the original article will probably have been summarised in a journal called *Chemical Abstracts*. Look the trade name up in the cumulative subject index to the abstracts (each volume covers four years); if your product is listed the index will tell you which issue of *Chemical Abstracts* summarised the article. (A fuller description of how to use abstract journals is given in Part 6 of this chapter.)

The least painful—but most expensive—do-it-yourself way of identifying the composition of trade-name products is to use a computerised chemical dictionary called *CHEMLINE* which lists the ingredients of many British trade-name products. You do not need any special training or knowledge of computers to use this system—but you do need money. It costs a little over £25 (at 1977 prices) to consult *CHEMLINE*, so it may only be worth doing if you have a long list of products to identify. More details are given on pages 40–1.

Part 4—Organisations that may help you with information

Before you put time into searching through libraries for information on chemical hazards, see whether you can use the law to help you get the information you need—or whether any organisation with experience in the subject can supply you with answers to your questions.

Your employer

If you are enquiring about a substance used in your place of work, your employer is required by law to keep you informed about its hazards. Under the 1974 Health and Safety at Work Act it is the duty of every employer to provide his employees with: 'such information ... as is necessary to ensure, so far as is reasonably practicable, the health and safety at work of his employees' *(section 2(2)(c))*.

Your employer should give you details of any known or suspected hazards of the substances you work with and of any monitoring he has done to see what concentrations are present in the workplace air.

If you work with raw materials that have been manufactured, imported or supplied to your employer by another company, it is the duty of that company to make sure that (a) they can be used without risk to health, (b) they have been properly tested, and (c) information about any health hazards and the results of any tests are available (Health and Safety at Work Act, *section 6(4)*). If you ask for these details, your employer must pass them on to you or to your trade union or safety representative.

This procedure may give you all the information you need. ICI, who manufacture 1,1,1-trichloroethane under the trade name Genklene, provide their customers with a small pamphlet entitled 'The Safe Use of "Genklene" '. This pamphlet stresses Genklene's advantages over other solvents (less toxic and inflammable), advises on safe methods of using and disposing of the solvent, summarises its hazards, states the TLV and recommends first aid procedures in case of overexposure.

All the hazards discussed in the main reference books on toxic substances are mentioned in the ICI pamphlet, although in less detail than is given in figure 2.1, for example. However, the pamphlet does not state the concentrations at which ill-effects have been found. You would have to look to other published sources for this information.

The Factory Inspectorate

If your employer seems unwilling or unable to supply you with full details of the hazards of substances you work with, contact the local offices of the Factory Inspectorate (who may be listed under Health and Safety Executive in the telephone directory). The Inspectorate is required by law to help keep employees or their representatives 'adequately informed about matters affecting their health, safety and welfare'. The Inspector is required to give you (a) factual information that he receives about working conditions that may be harmful, and (b) details of action that he has taken or proposes to take to deal with conditions he finds in your workplace (Health and Safety at Work Act, *section 28(8)*).

The chemical manufacturer

If you do not get all the information you need from your employer or the Factory Inspectorate, you can try writing directly to the manufacturer of the substance. You may have to do this for information about a substance that is not used in your own place of work. The manufacturer is not required by law to supply information about the hazards of his products to anyone other than his own customers, though some manufacturers say they will provide this information to anyone asking for it.

The manufacturer's name and address should be printed on the container of the substance. Give a full description of the substance—including any code markings that may refer to its contents—and ask for details of:

(1) The proper chemical names of *all* the ingredients in the product, if it is one that is marketed under a trade name. Ask for the proportions in which the different chemicals are mixed.

(2) *All* reports of the known or suspected hazards of the substances including details of the concentrations used in any tests and a full description of any ill-effects found.

(3) Recommended precautions for using and disposing of the substances.

(4) First aid or emergency measures to be used in case of over-exposure or accidental release of the substances.

Trade union services

If you are a member of a trade union, you can ask for help from your union's research department. Trade unions may pass enquiries they cannot deal with themselves on to the Trades Union Congress medical adviser who may visit the factory and arrange for exposed workers to be examined. The union may also pass the question on to the TUC Centenary Institute of Occupational Health for special advice.

A request for information through official trade union channels can take several weeks, and you may often have to wait for months. Trade union members may be able to get a quick answer to a problem by telephoning the TUC Centenary Institute directly (telephone: 01 636 8636, extension 392) and talking to the Information Officer.

Other bodies

The British Society for Social Responsibility in Science (9 Poland Street, London W1, telephone: 01 437 2728) helps trade unionists or tenants' and residents' associations look into the hazards of toxic substances found at work or in the environment. There are few other bodies who specifically state their willingness to provide information to enquirers on environmental hazards. Try the official pollution control authorities: the Alkali Inspectorate and local authorities' Environmental Health Departments for air pollution, the water authorities for river pollution (the Water Research Centre—see page 56—can also be helpful), or the Waste Disposal Departments of County Councils.

Part 5—How to use published sources of information on toxic hazards

Use this section to help you look up the hazards of a toxic substance used in the workplace or found in the environment. If you are interested in the effects of an air pollutant or a water pollutant, start with Part 8 or Part 9 of this chapter and then return to this section if you need further details.

Several different reference books on hazards are described here: some of them are more detailed or more complicated to use than others. To help you decide for yourself which is best suited for your purposes, their entries for 1,1,1-trichloroethane have been reproduced in figures 2.1–2.7.

Although many of the books are published in the United States, a good science library in this country should hold at least some of them. Nearly all the sources listed in this chapter are on the shelves of the British Library's Science Reference Library in London at either the Holborn (telephone: 01 405 8721) or Bayswater (telephone: 01 727 3022) branches.

Your local library will be able to borrow any of these books for you from the British Library's Lending Division. If you think you will need to consult one of these books regularly, ask your library to buy a copy. If you work for a company that uses a variety of chemicals, and you have access to the company's library, ask its library to buy one of these books. If, failing all else you decide to buy one of these books yourself (and some cost as much as £40) you can order it yourself from the publisher or ask a local bookshop to get it for you. Details of how to order US government publications are given in Appendix 2.

THE CODES
We have put a simple code before the title of each book in this section: it tells you how detailed the information given in the book is and, if

possible, how many different substances are covered by the book.

A FULL summary of information includes details of all the following:

(1) the chemical formula of the substance and its alternative names (synonyms);

(2) a description of its known toxic effects including the concentration and route of exposure (for example, inhaling, swallowing etc.) responsible for any effect;

(3) the Threshold Limit Value, if one has been set for the substance (but see chapter 1 for criticism of TLV standards);

(4) references to the original reports describing the toxic effects. These will help you to look up a report for yourself if you need more details than the book contains.

A BRIEF summary leaves out at least one of these items of information; on its own, it will probably not give you all the details you need.

More recent editions of these books may have been published since this Handbook was written: it may be worth checking this with a library or bookshop. Remember to use the most recently published of the sources available to you, and to use one of the updating sources described in Part 6 if your source is not recent.

General reference books

FULL/Number of substances not stated

Encyclopaedia of Occupational Health and Safety 1972. International Labour Office. Geneva, Two volumes, A–K and L–Z. 1621 pages. £40. Available from the ILO's London office, 87/91 New Bond Street, London W1.

For those substances it covers, the ILO's *Encyclopaedia* is one of the best single sources of information on hazards (see figure 2.1). The two volumes deal with toxic hazards and many other aspects of occupational health and safety in articles written 'for the benefit of all persons concerned with workers' protection. . . . It is not intended exclusively for specialists'.

FULL/500 substances

Documentation of the Threshold Limit Values for Substances in Workroom Air, American Conference of Governmental Industrial Hygienists (1971, 3rd printing 1976) (includes a supplement for changes in

TLVs between 1971 and 1975). $15.00. Available from ACGIH, PO Box 1937, Cincinnati, Ohio 45201.

The *Documentation* contains the evidence on which the Threshold Limit Values (TLVs) for about 500 different substances have been set. It gives a more detailed summary of reports of hazards than any of the other books in this section and provides a guide to the degree of protection likely from any TLV (see figure 2.2). The book describes the most recent evidence available at the time that any TLV was set: it may be relatively out-of-date where substances whose TLVs have not been changed for some time are concerned.

BRIEF/16 500 substances in 1975 edition

Registry of Toxic Effects of Chemical Substances. (Formerly known as 'Toxic Substances List'), National Institute for Occupational Safety and Health, US Department of Health, Education and Welfare. Published annually. 1975 edition: 1296 pages, $9.65. Report No. 1733–00096. Available from Superintendent of Documents, US Government Printing Office, Washington DC 20402 (see page 172).

The *Registry* has several advantages over other books in this section. (1) It covers more toxic substances than any other reference book on the subject; (2) the alphabetical list contains synonyms and trade names, where these refer to single substances; and (3) it is updated every year, and therefore includes the most recent findings.

Unfortunately, the *Registry* is rather difficult to use as its entries are almost entirely in the form of codes or abbreviations (see figure 2.3), though these symbols are all explained in the text.

It has two other drawbacks:

(1) Most of the substances included in editions of the *Registry* published up till 1975 were chosen because their lethal dose in animals was known. Many substances known only to produce non-lethal effects were not covered, though they may be included in later editions.

(2) Where non-lethal effects are shown, the description states only which part of the body is affected—but not what the effects are.

Thus, although the Registry contains information that may not be found in other reference books, *you should never try to use it as your only source of information on a toxic substance.*

BRIEF/13 000 substances

Dangerous Properties of Industrial Materials, N.I. Sax. (New York: Van Nostrand Reinhold, 4th Edition 1975). 1258 pages.

This is another book which covers a very large number of chemicals

Investigating Hazards

with very abbreviated information. It relies on a coded 'toxic hazard rating' which is fully explained in the text, but which does not give a proper description of what the effects are. Many of the entries—like the one shown in figure 2.4—are extremely brief, though some substances are given a more useful written account. Entries make it clear that the absence of information on a substance does not mean that it is harmless but rather that its toxic properties have not been adequately investigated. The book covers many chemicals that are not dealt with elsewhere but may not, on its own, give you all the information you will need.

BRIEF/1300 substances

Clinical Toxicology of Commercial Products, R. E. Gosselin and others (previous editions edited by M. Gleason). (Baltimore: Williams and Wilkins Co., 4th Edition 1976). $54.00.

This book deals mainly with the hazards of commercial brand-name products sold in the USA which might be hazardous if swallowed or inhaled accidentally. It also lists typical formulae for common products, pointing out in each case which of the ingredients is likely to be the most dangerous.

The book includes a section on the toxic hazards of 1300 substances used as ingredients in commercial products; its information is equally useful as an account of the hazards to workers who handle these chemicals (see figure 2.5).

BRIEF/100 substances

Environmental and Industrial Health Hazards: A Practical Guide, R. A. Trevethick. (London: William Heinemann Medical Books, 1976). 214 pages, £5.75.

This is one of the few guides to toxic hazards that has actually been written for the lay person: unfortunately it only covers a relatively small number of the most common industrial substances. Each substance is covered by two pages, one on its hazards, the other on medical data. The sheets are designed to be displayed in places of work so that everyone concerned can see the precautions that should be taken when handling the substance, the symptoms of acute and chronic poisoning, recommended first-aid measures and medical treatment, and Factory Act regulations that apply to the substance. However, the book does not state the concentrations known to produce the ill-effects it describes (see figure 2.6).

Trichloroethanes

HAZARDS

Absorption of a substantial amount of 1,1,1-trichloroethane from the gastrointestinal tract will produce a functional depression of the central nervous system. If the amount ingested is sufficient to cause unconsciousness, impairment of liver function may result. Several drops of 1,1,1-trichloroethane splashed directly on the cornea can result in a mild conjunctivitis which will subside within a few days. Prolonged or repeated contact with the skin results in transient erythema and slight irritation, owing to the defatting action of the solvent.

1,1,1-trichloroethane is rapidly absorbed through the lungs and the intestinal tract. Following absorption, most of the compound is eliminated unchanged via the lungs. A small percentage is metabolised to carbon dioxide, while the remainder appears in the urine as the glucuronide of 2,2,2-trichloroethanol. The absorption of toxic amounts of 1,1,1-trichloro-ethane through the skin in the course of normal industrial operations is highly unlikely, unless the solvent is confined to the skin beneath an impermeable barrier.

Judging from human toxicological data, 1,1,1-trichloroethane has proved to be one of the least toxic of the chlorinated aliphatic hydrocarbon solvents.

Acute exposure. The principal toxic action of a single vapour exposure is a depression of the central nervous system, proportional to the magnitude of exposure, and typical of an anaesthetic agent. Humans exposed to 900–1 000 ppm experienced transient, mild eye irritation and prompt, though minimal, impairment of co-ordination. Above 1 700 ppm, obvious disturbances of equilibrium have been observed. Exposures of this magnitude may also induce headache and lassitude, while nausea has not been reported. Below the current TLV, no physiological effects have been observed. Human

---

BRIEF/Number of substances not stated

Poisoning by Drugs and Chemicals. An Index of Toxic Effects and Their Treatment, Peter Cooper. (London: Alchemist Publications, 3rd Edition 1974). 218 pages, £2.85.

This is an interesting source of information on case histories of poisoning caused by drugs and chemicals. It contains more than most of the other sources on individual cases of overexposure, but its labelling of chemicals is imprecise. The entry shown in figure 2.7 does not make it clear that it refers to 1,1,1-trichloroethane and not to the 1,1,2-isomer. Confusing these two compounds could be very misleading.

experiments demonstrated a variation in human response to a given concentration of 1,1,1-trichloroethane vapour. Of the test procedures investigated, the modified Romberg test (subject balances on one foot, with eyes closed and arms at his sides) has proved to be the most sensitive, objective neurological sign of over-exposure.

Repeated exposure. It is unlikely that significant organic injury will result from repeated vapour exposure in the absence of acute anaesthetic effects. No injury to man following repeated exposures to vapour concentrations of less than 500 ppm has been observed.

*TORKELSON, T. R.; OYEN, F.; McCOLLISTER, D. D.; ROWE, V. K. (1958). Toxicity of 1,1,1-trichloroethane as determined on laboratory animals and human subjects. *American Industrial Hygiene Association Journal,* **19,** 353.

*STEWART, R. D.; GAY, H. H.; ERLEY, D. S.; HAKE, C. L.; SCHAFFER, A. W. (1961). Human exposure to 1,1,1-trichloroethane vapor: relationship of expired air and blood concentrations to exposure and toxicity. *American Industrial Hygiene Association Journal,* **22,** 252.

*STEWART, R. D.; DODD, H. C. (1964). Absorption of carbon tetrachloride, trichloroethylene, tetrachloroethylene, methylene chloride, and 1,1,1-trichloroethane through the human skin. *American Industrial Hygiene Association Journal,* **25,** 439.

*STEWART, R. D.; GAY, H. H.; SCHAFFER, A. W.; ERLEY, D. S.; ROWE, V. K. (1969). Experimental human exposure to methyl chloroform vapor. *Archives of Environmental Health,* **19,** 467.

FIGURE 2.1

Extract from *Encylopaedia of Occupational Health and Safety* (1972). (Only part of the entry on trichloroethanes is shown: the remainder deals with the production and uses of the substances and advice on the diagnosis and treatment of trichloroethane poisoning)

Guides to individual substances

Some slightly more specialised reference books deal with the toxic hazards of just one class of substances, for example, solvents, adhesives or carcinogens. Lists giving references to such books have been produced by the TUC Centenary Institute for Occupational Health (see page 25) and by the Science Reference Library.[3]

Guides to the hazards of a number of individual substances are produced by some organisations; as well as listing hazards, these guides explain the precautions that should be used when handling the material and advise on first-aid and medical treatment in cases of overexposure.

METHYL CHLOROFORM (1,1,1-Trichloroethane)
CH_3CCl_3

350 ppm (Approximately 1,900 mg/m³)

Torkelson and associates(1) described the toxicity of methyl chloroform from repeated exposures of animals and single exposures of men. Exposure of animals for three months at concentrations from 1000 to 10,000 ppm caused some pathologic changes in the livers and lungs of some species; the main effect of exposure appeared to be anesthesia. Exposure to the vapor at 500 ppm for seven hours a day, five days a week, for six months did not cause any toxic changes of significance in rats, guinea pigs, rabbits, or monkeys. Exposures of men at 920 ppm for 70 minutes resulted in equivocal evidence of readily reversible toxicity; however, exposures at 1900 ppm for five minutes caused a disturbed equilibrium in exposed men.

According to Stewart(2) methyl chloroform is the least hepatotoxic of the common chlorinated hydrocarbon solvents (methylene chloride not included in comparison). No injury to man following repeated exposure at concentrations below 500 ppm has been observed. Men exposed at 900 to 1000 ppm experienced transient mild irritation and minimal impairment of coordination. Above 1700 ppm obvious disturbance of equilibrium has been observed. Exposures of this magnitude may induce headache and lassitude, but nausea was not reported.

Rowe and associates(3) found that the only effect of repeated exposure of several species at 500 ppm of a mixture containing 75% methyl chloroform and 25% perchloroethylene was a slight degression in the growth of guinea pigs, due to a reduced food intake. At 1000 ppm mild reversible liver and kidney changes were detected. A time-weighted average limit of 400 ppm was recommended for this mixture. Stewart et al.(4) studied the effects of exposure of human subjects at 500 ppm for seven hours a day for five days. Subjective responses were mild (sleepy feeling) and of doubtful significance. The only adverse objective response was an abnormal modified Romberg test in two of eleven subjects.

Since its introduction as a solvent, methyl chloroform has found wide application and few reports of serious ill effects have been recorded. Hatfield and Maykoski(5) reported a death resulting from the cleaning of a 470-gallon tank. An hour and twenty minutes later the air in the tank contained 500 ppm of methyl chloroform, but reconstruction of the accident resulted in concentrations of 36,000 to 62,000 ppm. Reports of six other fatalities, four from work inside tanks were noted. Hake et al.(6) studied the metabolism of methyl chloroform, and found that over 98% was excreted in the breath. They suggested that this inertness in the body may be a reason for the low toxicity.

Patty(7) and others have suggested 500 ppm as a TLV. Because of complaints due to odor and mild irritation during use(8), and concern over the possible effects of prolonged undue exposure to chlorinated hydrocarbons(9), a TLV of 350 ppm is recommended. ANSI recommended 400 ppm in 1970.

References:
1. Torkelson, T.R., Oyen, F., McCollister, D.D., Rowe, V.K.: Am. Ind. Hyg. Assn. J. *19*, 353 (1958).
2. Stewart, R.D.: J. Occ. Med. *5*, 259 (1963).
3. Rowe, V.K., Wujowski, T., Wolf, M.A., Sadek, S.E., Stewart, R.D.: Am. Ind. Hyg. Assn. J. *24*, 541 (1963).
4. Stewart, R.D., Gay, H.H., Schaffer, A.W., Erley, D.S., Rowe, V.K.: Arch. Env. Health *19*, 467 (1969).
5. Hatfield, T.R., Maykoski, R.T.: Arch. Env. Health *20*, 279 (1970).
6. Hake, C.L., Waggoner, T.B., Robertson, D.N., Rowe, V.K.: Arch, Env. Health *1*, 101 (1960).
7. Patty, F.A.: Industrial Hygiene and Toxicology, Vol. II, 2nd Ed., p.1288, Interscience, N.Y. (1963).
8. Coleman, A., Stapor, J.: Communication to TLV Committee (1962).
9. Elkins, H.B., Stokinger, H.E.: Communication to TLV Committee (1962).

FIGURE 2.2
Extract from *Documentation of the Threshold Limit Values* (1971)

Investigating Hazards 33

KJ29750[1] **ETHANE, 1, 1, 1-TRICHLORO-**
  CAS:[2] 000071556   MW:[3] 133.37   MOLFM:[4] Cl3-C2-H3
  WLN:[5] GXGG
  SYN:[6] AEROTHENE TT   *   CHLOROFORM, METHYL-   *
      CHLOROTHENE (Inhibited)   *   CHLORTEN   *
      METHYLCHLOROFORM   *   1, 1, 1-TRICHLOOETHAAN
      (Dutch)   *   1, 1, 1-TRICHLORAETHAN (German)   *
      alpha-TRICHLOROETHANE   *   1, 1, 1-TRICHLOROETHANE   *
      1, 1, 1-TRICLOROETANO (Italian)   *
  TXDS:[7] ihl-man[8] TCLo:[9] 350 ppm TFX:PSY[10]   WEHL** 10,82,73[11]
      ihl-hmn[12] TCLo: 920 ppm/70M[13] TFX:CNS[14] AIHAAP 19,353,58[15]
      orl-rbt[16] LD50 :5660 mg/kg[17]   AIHAAP 19,353,58
      orl-gpg[18] LD50 :9470 mg/kg   AIHAAP 19,353,58
  AQTX:[19]
  U.S. OCCUPATIONAL STANDARD USOS-air[20]   FEREAC 37,22139,72[21]
  TWA 350 ppm[22]

Notes
(1) The *Registry's* code number for the substance.
(2) The *Chemical Abstracts Registry* number. You will not need to use this in a normal search.
(3) The molecular weight.
(4) The molecular formula.
(5) The Wiswesser Line Notation. A code for use with computer searches. You will not need this in a normal search.
(6) Synonyms: other chemical names, including those used in other languages, trade names and common names. All the synonyms listed here are also included in the main alphabetical list of entries. If you looked one up, the entry would instruct you to look up ETHANE, 1,1,1-TRICHLORO-, that is, the entry shown here.
(7) Toxic Dose Data.
(8) The route of administration and the species used. In this case, inhalation and man.
(9) The lowest published toxic concentration, 350 ppm.
(10) 'TFX' stands for 'toxic effects'. In this case the effects were 'psychotropic' ('PSY'), that is, psychological or behavioural.
(11) The reference to the journal in which the original report of this study was published. In this case *Work-Environmental Health* (Stockholm). Vol. 10, page 82 (1973). This same report is also described in the first abstract shown in Figure 2.9.
(12) Route of administration—inhalation, species—human.
(13) Period of exposure, 70 minutes.
(14) Toxic effects involved the central nervous system ('CNS').
(15) The reference. *American Industrial Hygiene Association Journal* Vol. 19, page 353 (1958).
(16) Route of administration—oral; species—rabbit.
(17) The dose needed to kill 50 per cent of test animals ('Lethal Dose 50') was 5660 milligrams of substance per kilogram of body weight.
(18) Route of administration—oral; species—guinea pig.
(19) Information on the aquatic toxicity ('AQTX') of the substance to certain species of water organisms is available and shown in another section of the *Registry*.
(20) A standard limiting the amount of this substance permitted in the air of workplaces has been set under the US Occupational Safety and Health Act of 1970.
(21) The reference to the location of the standard in the *Federal Register*, a publication containing US government regulations and official documents.
(22) The US standard for this substance in workplace air is a time weighted average ('TWA') of 350 ppm.

FIGURE 2.3
Extract from *Registry of Toxic Effects of Chemical Substances* (1975 edition)

α-TRICHLOROETHANE *
General Information
Synonyms: 1,1,1-trichloroethane; methyl chloroform.
Colorless liquid.
Formula: CH_3CCl_3.
Mol wt: 133.42, bp: 74.1°C, fp: −32.5°C, flash p.: none, d: 1.3492 at 20°/4°C, vap. press.: 100 mm at 20.0°C.
Hazard Analysis
Toxic Hazard Rating:
 Acute Local: Irritant 1; Ingestion 1.
 Acute Systemic: Inhalation 2.
 Chronic Local: Irritant 1.
 Chronic Systemic: Ingestion 1; Inhalation 1.
Toxicity: Narcotic in high concentrations.
Disaster Hazard: Dangerous; See chlorides.
Countermeasures
Ventilation Control: Section 2.
Personal Hygiene: Section 2.
Storage and Handling: Section 7.
Shipping Regulations: Section 11.
Regulated by IATA.

* This material has been assigned a Threshold Limit Value by ACGIH. See complete reprint of TLV's in Section 1.

TOXIC HAZARD RATING CODE (For detailed discussion, see Section 9.)

0 NONE: (a) No harm under any conditions; (b) Harmful only under unusual conditions or overwhelming dosage.

1 SLIGHT: Causes readily reversible changes which disappear after end of exposure.

2 MODERATE: May involve both irreversible and reversible changes not severe enough to cause death or permanent injury.

3 HIGH: May cause death or permanent injury after very short exposure to small quantities.

U UNKNOWN: No information on humans considered valid by authors.

FIGURE 2.4
Extract from *Dangerous Properties of Industrial Materials* (1975)

1,1,1-TRICHLORO-
ETHANE
Alpha-trichloroethane
Methyl chloroform
Chlorothene

3 Widely used as a solvent. Absorbed through the lungs, but toxic effects occur only with concentrations more than 500 p.p.m. There have been two reported human deaths from exposure to high vapor concentrations in unventilated drums. Single oral toxicity is low. Skin contact causes only mild irritation even on prolonged or repeated exposure. Systemic toxicity is approximately that of methylene chloride and considerably less than carbon tetrachloride. Animal tests place it in toxicity class 2, but 3 may be a better estimate for man. Because like many solvents, 1, 1, 1–trichloroethane sensitizes the heart to epinephrine, every precaution should be taken not to increase circulating epinephrine. (See also: CARBON TETRACHLORIDE, Reference Congener in Section III.) Ref.: von Oettingen, 1955; Torkelson et al., 1958.

FIGURE 2.5
Extract from *Clinical Toxicology of Commercial Products* (1969 edition)

Trichloroethane $C_2H_3Cl_3$

Synonyms Methyl Chloroform
Vinyl Trichloride

Trichloroethane $C_2H_3Cl_3$
Threshold Limit Value: 1,1,1 Isomer 350 p.p.m. 1900 mg./m³
1,1,2 Isomer (Skin) 10 p.p.m. 45 mg./m³.

Hazard Data Sheet

General Data – Grade II Chemical
It exists as 2 isomers 1,1,1 and 1,1,2 Trichloroethane. Authorities are contradictory as to which is the more toxic of the two isomers. The values attached to the threshold limit value suggest that 1,1,1 is less toxic. It is a colourless liquid with a characteristic odour of chloroform.

Uses
The 1,1,1 isomer is used as a solvent for resins, oils, waxes and tar and is a substitute for carbon tetrachloride. It is a degreasing agent.
The 1,1,2 isomer has very little industrial use. For the purposes of this data sheet it will be disregarded except to state that its toxic symptoms are similar to the 1,1,1 isomer.

Hazard
Trichloroethane possesses anaesthetic properties leading to drowsiness, unconsciousness, respiratory failure, and death. Its effect is accentuated if used in confined spaces.
It is a powerful defatting agent causing dry and cracking skin.

Precautions
Where used as a degreasing agent the tank should have a condensing coil around the mouth of the tank and there should be forced draft ventilation with lip extraction.
Protective clothing i.e. plastic gloves and goggles may need to be supplied to operatives in close contact with the process. Removal of contaminated clothing if splashes have occurred.

Acute Poisoning
1) Loss of co-ordination and balance.
2) Mild discomfort of eyes and nose.
3) Headache and lassitude.
4) Unconsciousness leading to death if exposure is continued. If the liquid is swallowed the effects will be similar to those described above though vomiting and diarrhoea may also occur.

Chronic Poisoning
Recovery from the acute episode is usually complete. There are no long term effects apart from dry cracking skin from the defatting action.

First Aid Measures
1) Remove from the hazardous area.
2) Apply artificial respiration if needed together with oxygen therapy.
3) Remove contaminated clothing and wash skin.
4) Send for medical attention.

Factories Act References and Regulations
S (4)
S (30)
S (63)

FIGURE 2.6
Extract from *Environmental and Industrial Health Hazards: A Practical Guide* (1974)

TRICHLOROETHANE
Synonyms: methylchloroform; 'Chlorothene'; 'Genklene'; in 'Thawpit'.
Occurrence: Dry-cleaning fluids.
Action: Narcotic.
Absorption and Excretion: Readily absorbed via the lungs or gastrointestinal tract and largely excreted by the lungs.
Toxic Effects: Inhalation depresses central nervous functions and leads to respiratory or cardiovascular failure. Kidney and liver dysfunction are minimal and likely to be transient. Exposure to 960 to 1000 ppm of vapour causes transient eye irritation and impairs nervous coordination. Equilibrium is upset at about 1700 ppm. Headache and lassitude occur, but nausea is rare except after ingestion. Sensitisation of the myocardium to pressor catecholamines is a hazard of exposure. TLV 350 ppm.

A man aged 27 was found unconscious, cyanosed and apnoeic in an aircraft fuel tank which he had been cleaning with trichloroethane. He died in spite of attempts at artificial ventilation. Passive congestion of lungs and kidneys, and pulmonary œdema, appeared at post mortem examination. Cerebral œdema was slight. The blood content of trichloroethane was 6 mg per 100 ml. Experiments showed that up to 62,000 ppm of solvent vapour could be produced if swabbing pads were used inside the tank. This represents more than double the anæsthetic concentration. (T. R. Hatfield and R. J. Maykoski, *Archs envir. Hlth*, 1970, **20**, 279.)

A man aged 47 accidentally swallowed 30 ml of liquid believed to be whisky but actually trichloroethane. He had a burning sensation of mouth and upper alimentary tract, and half an hour later felt nauseated. One hour later he started vomiting and suffered from diarrhœa, both becoming more severe over the next 90 minutes. He was given gastric lavage with two litres of tap water. He was orientated, coordinated and not drowsy, and showed no neurological abnormalities. Six hours after the accident the vomiting and diarrhœa subsided and the patient grew fatigued and drowsy. Next day he awoke free from symptoms but was kept under observation for two weeks. Much trichloroethane was excreted in the patient's breath. (R. D. Stewart and J. T. Andrews, *J. Am. med. Ass.*, 1966, **195**, 904.)
Exposure to more than 1000 ppm of trichloroethane for 15 minutes, or to 2000 ppm for five minutes may disturb equilibrium in most adults. Five minutes' exposure to 5000 to 10,000 ppm produces marked nervous incoordination and anæsthesia. Ingestion causes nausea, vomiting and diarrhœa. The estimated lethal dose is 8·6 to 14·3 g per kg body-weight. Breath analysis by infra-red spectrography affords the best means of determining the degree of exposure to vapour. (R. D. Stewart, *J. Am. med. Ass.*, 1971, **215**, 1789.)

Suggested Treatment: As described under trichloroethylene (p. 208).
Aids to Identification: A colourless liquid boiling at about 74° and insoluble in water; miscible with ethanol, ether or acetone. With strong sodium hydroxide solution and pyridine, warmed on a boiling water bath, it produces a red colour in the pyridine layer (not specific).

FIGURE 2.7
Extract from *Poisoning By Drugs and Chemicals: An Index of Toxic Effects and Their Treatment* (1974)

HEALTH AND SAFETY EXECUTIVE GUIDANCE NOTES

The UK Health and Safety Executive produces a set of notes (formerly known as 'Technical Data Notes') on about 50 different subjects—about half of them covering the health hazards of specific substances. The Notes are available from the Health and Safety Executive head office (Baynards House, 1, Chepstow Place, London W2, telephone: 01 229 3456), from any of its local offices or from HMSO.

NIOSH CRITERIA DOCUMENTS

When the US National Institute for Occupational Safety and Health recommends a new occupational standard for a toxic substance, it publishes a 'criteria document', which may run to more than 100 pages, giving, amongst other things, a very complete summary of the reported toxic effects of the substance in humans and animals, and full instructions on safe methods of handling the substance. A list of all NIOSH publications can be obtained from NIOSH, Post Office Building, Room 530, Cincinnati, Ohio 45202, USA.

CHEMICAL SAFETY DATA SHEETS

These notes are produced by the American Manufacturing Chemists Association and are similar to the Health and Safety Executive's Guidance Notes, though they cover many more substances and give a more detailed account of the health hazards and recommended safety precautions. A list of the substances covered by these data sheets can be obtained from the AMCA, 1825 Connecticut Avenue NW, Washington, DC 20009, USA.

Part 6—Bringing your information up-to-date

If you have looked up the hazards of a substance in a book that is more than, say, two to three years old, you may need to know if there is any more recent news about it. The older your main source of information is, the more important it becomes to update it.

Using 'updating sources' may also be the only way to find out if anything is known about the hazards of the many thousands of different chemicals that are not mentioned at all in most reference books.

There are two sources of updating information: abstracts journals and the *TOXLINE* computer system.

Abstract journals

Each month, new reports on the hazards of materials are published in a variety of different scientific journals. To save you looking through all these journals, special journals summarising these reports in the form of short 'abstracts' are published.

At the end of each year, each abstracts journal produces an index to the titles of the reports it has published on each toxic substance, or other subject, during the year.

If you had used a reference book published in 1973 to look up the hazards of 1,1,1-trichloroethane and you wanted to know if any new hazards had been discovered since then, you could look the substance up in the annual indexes from 1974 onwards of an abstracting journal such as *CIS Abstracts* published by the International Labour Office. Figure 2.8 is taken from the 1974 index of *CIS Abstracts* and shows the titles of articles published during 1974 on the health hazards of trichloroethane.

TRICHLOROETHANE
Trichloroethane concentration in alveolar air and blood at rest and during exercise   CIS 74-443

Effect of trichloroethane on psychophysiological functions   CIS 74-444

Cardiovascular effects of 1,1,1-trichloroethane
    CIS 74-1675

Resistance of protective gloves to trichloroethane
    CIS 74-2001

FIGURE 2.8
Extract from the 1974 index to *CIS Abstracts*

The title of each article in the index is followed by a code number which tells where, in the regular issues of *CIS Abstracts*, the summary of this article appeared. Two of the summaries (abstracts) listed in the index are shown in figure 2.9.

Sometimes, you will find all the information you need about a new report contained in the abstract itself. For example, the top abstract shown in figure 2.9 tells you that specific ill-effects occurred at a particular concentration of 1,1,1-trichloroethane. (Note that this report is the same one described in figure 2.3—notes 7 to 11—as giving the lowest concentration at which ill-effects had been reported in human beings.) In other cases—for example, the second abstract in figure 2.9—the abstract does not tell you the results of the experiment it describes. If you want to know what the results were, you will have to read the original article itself.

> **CIS 74-444 Methylchloroform exposure – II. Psychophysiological functions.** Gamberale F., Hultengren M. *Work – Environment – Health*, Helsinki, Finland, 1973, Vol.10, No.2, p.82-92. Illus. 23 ref. (In English)
>
> Reaction time, perceptual speed and manual dexterity were studied in 12 healthy men repeatedly exposed to 250, 350, 450 and 550 ppm of 1,1,1-trichloroethane (TCE) in inspired air; control tests were carried out with pure air. Exhaled air samples were taken at 2-min intervals. There was a linear relationship between the concentrations in alveolar air and arterial blood. Reaction time, perceptual speed and manual dexterity were impaired during exposure to TCE as compared with exposure to pure air. Analysis of the results shows that psychophysiological functions are adversely affected by exposure to a TCE vapour concentration of 350 ppm. (00376)

> **CIS 74-2001 Resistance of protective gloves to industrial solvents – Results obtained with trichloroethane on 100 commercial gloves** (Résistance des gants de protection aux solvents industriels – Résultats obtenus avec le trichloréthane sur une centaine de gants du commerce). Chéron J. *Travail et sécurité*, Paris, France, Oct. 1973, No.10, p.493-498. Illus. (In French)
>
> Presents the results of tests carried out by the French National Research and Safety Institute on the deterioration of gloves by soaking them into the solvent (combined mechanical and chemical actions), permeability to the solvent and speed of solvent passage across the glove. The findings are set out in a table with a qualitative scale, graded according to how the gloves are used (prolonged contact, frequent dipping, etc.) and accompanied by advice to users. (01904)

FIGURE 2.9
Two entries from *CIS Abstracts* in 1974

The full reference to the original article shown in the abstract is given after the title. The name of the publication is in italics, followed by the volume, issue and page numbers and the language in which the article is written.

Summaries of, or references to, new reports on the health hazards of industrial chemicals and environmental pollutants can be found in the following sources:

Abstracts on Hygiene
Carcinogenesis Abstracts
CIS Abstracts
Cumulated Index Medicus
Current Bibliography of Epidemiology
Excerpta Medica
   Section 2C—Pharmacology and Toxicology
   Section 16—Cancer
   Section 17—Public Health, Social Medicine and Hygiene
   Section 35—Occupational Health and Industrial Medicine

Index Medicus
Industrial Hygiene Digest
Occupational Safety and Health Abstracts
Toxicity Bibliography

Abstracting journals that deal specifically with air and water pollution are shown on pages 49 and 57.

Computer searches

The easiest, but most expensive, way of digging up recent references on a substance is to ask a computer linked up to a computerised 'file' of information on toxicity. Do not be put off by the thought of having to deal with a computer: you will not have to operate it yourself—you simply explain what you are looking for to a trained computer operator and he or she will do all the work. If you phrase your question precisely, the computer will print out for you a list of all reports on the toxicity of your substance published between 1930 and last month.

Information on the health hazards of industrial chemicals, environmental pollutants and pesticides is stored on a file called *TOXLINE* which contains over 400 000 references to reports published since 1971. Earlier reports are covered by *TOXBACK*, which deals with the period between 1930 and 1971.

A computerised chemical dictionary called *CHEMLINE* is also available. *CHEMLINE* lists all the known synonyms and many of the trade names of the substances covered in *TOXLINE*. A *CHEMLINE* search may help you to identify the chemical ingredients of trade-name products.

There are two ways of carrying out a computer search. You can have a search done for you by writing to the British Library, Lending Division, Medlars Section, Boston Spa, Wetherby, Yorkshire. It may help if you telephone them first (0937 843434) and ask for their advice on how to phrase the question in your letter.

Questions have to be very precise. If you simply asked for 'everything you have on trichloroethane' you would probably receive a mass of largely irrelevant information. You would need to make it clear that you were interested in (a) health hazards to man including the results of animal experiments of (b) exposure to 1,1,1-trichloroethane, (c) by any route of absorption (or by a specific route such as inhalation) and that you wanted (d) either references or abstracts to (e) any articles published since a particular date.

You can ask for information on up to three different substances in a single search. The cost of a *TOXLINE* and *TOXBACK* search is £25 (at 1977 prices) plus about 1½ p per reference. The British Library will supply you with up to 100 references, if they find that many. If you want to have a summary of each article printed out for you, ask them to send 'abstracts'—each one will cost about 10p extra. The results of the search will be posted to you and will probably take between one and two weeks to arrive.

If you want to know the hazards of a substance that is only identified by a trade name or by one of its synonyms, you can have a combined *CHEMLINE* and *TOXLINE* search done for £40, plus a little extra for each reference. If you simply want to identify the chemical ingredients of a trade-named product—but do not want information about toxicity—you can have a *CHEMLINE* search on its own for £25 plus around 1p per reference. You should be able to have several different trade-name substances identified in a single search.

The alternative way of having a search done is to visit the computer terminal yourself, put your question and watch the answers as they are printed out. This gives you the chance to work by trial and error, changing your question slightly or adding new questions until you get all the information you need. The chances are that this method will give you better results than a postal search. You can carry out a search of this kind at the Bayswater branch of the Science Reference Library in London (10 Porchester Gardens, London W2, telephone 01 727 3022). However, the fees for a personal search are worked out on a different basis from the postal fees: you will be charged 60p *per minute* for using *TOXLINE* or *CHEMLINE* in this way, with a slight additional charge for each reference printed out.

Part 7—Make sure that pollution in your place of work is monitored

If you are looking at hazards in your own place of work, ask your employer for a list of ALL substances used there—not just those he thinks are dangerous.

In some cases it will be enough for you to show that a certain substance is particularly toxic, and on this basis to press for its replacement or for the use of stricter methods to control its release and protect you from exposure.

For other substances, the danger depends on the concentration. If the substance is in a form that can be inhaled, you will need to know what concentrations of it are actually present in the factory or workshop air, and how these concentrations compare to those known to be dangerous.

Make sure that your employer monitors the concentrations of any air contaminants and gives you the full results of all monitoring; the Health and Safety at Work Act requires him to provide you or your representatives with such information (see page 23).

The monitoring should:

(1) cover all the toxic substances likely to be in the air—whether in the form of gas, vapour, mist, fume or dust;

(2) be carried out in your presence so that you are satisfied that it covers those parts of the workplace where you actually work and where concentrations are likely to be greatest. The only way to get an accurate picture of the pollution concentrations to which you are exposed is to wear a 'personal sampler' on your body which measures the concentration of a substance in the air immediately around you;

(3) be carried out at times of the day when the substance monitored is in greatest use;

(4) is done regularly, because conditions change. Occasional monitoring will overlook shortlasting high concentrations which are potentially the most dangerous. Continuous monitoring, using automatic equipment, is the only way to get a full picture of air pollution concentrations (see pages 105–6).

If you are not happy with the monitoring done by your employer, contact the Factory Inspectorate. They may carry out their own monitoring and will show you the results if you ask for them.

You may also want to try carrying out your own monitoring. Advice on this, and many other aspects of health and safety at work is given in a very clearly written 'workers handbook' called *The Hazards of Work: How to Fight Them*.[4]

Insist that the concentrations of substances in the factory air are kept down below (a) the levels you have discovered are hazardous, (b) TLVs for the substances, if they have been set, and (c) are in any case as low as possible even if the hazards of a substance are completely unknown.

The Factory Inspectorate will give you a list of the most recent TLVs—but do not treat them as 'safe' levels. Chapter 1 explains the weaknesses of TLV limits.

When you look at monitoring results, bear in mind that all monitoring equipment has a certain inbuilt inaccuracy (see page 108). The method that is often used in factories involves a 'breathalyser' style tube of crystals that change colour if a certain chemical is present. These tubes will tell you whether a certain substance is in the air, but their readings may be as much as 30 per cent wide of the true concentration. If a reading obtained by this method shows that the concentration of a certain chemical in the air is 10 per cent below the TLV, the actual concentration may be 20 per cent above the TLV—or 40 per cent below it.

Part 8—Where to find information on air pollution hazards

This section tells you where to find information on the hazards of air pollution to the general population and the environment.

Most air pollution is caused by a relatively small number of common pollutants whose hazards are summarised in various general reference books; fuller details are given in more specialised reports on individual pollutants.

Health hazards

The following books contain fairly short summaries of the hazards of the commoner air pollutants:

Health Effects of Environmental Pollutants, George L. Waldbott. (The C. V. Mosby Company, St. Louis, 1973). (Separate chapters on pollutants according to the health effects they cause, such as pulmonary irritants, asphyxiants, allergens, carcinogens etc.)
Air Pollution. A Comprehensive Treatise: Vol. 1. Air Pollution and its Effects, Arthur C. Stern. (Academic Press, 3rd Edition, 1976). (Very brief details on the commoner pollutants)
Environmental Engineers Handbook Vol. 2. Air Pollution, Editor, Bela G. Liptak. (Radnor, Pennsylvania: Chilton Book Co., 1974). (Very brief descriptions of the hazards of pollutants, as well as their sources and any air quality standards)
Air Pollution and Health, Royal College of Physicians (London: Pitman Medical and Scientific Co Ltd., 1970). (A good, short book dealing mainly with the effects on human health of smoke and sulphur dioxide)

Rather more detailed accounts are given in:
Pollution Detection and Monitoring Handbook, Marshall Sittig (Noyes Data Corporation, 1974). (Contains information on nearly 90 air and water pollutants with a one to two-page summary on the hazards of each. Also contains much other useful information including details of air quality standards, typical sources of the pollutant,

US emission standards, and methods of measurement)
Industrial Pollution, N. I. Sax (New York: Van Nostrand Reinhold, 1975). (The last chapter in this very large book is in the form of an index, covering more than 100 pages, on the hazards of environmental pollutants)
Air Pollution and Cancer in Man: Proceedings of the Second Hanover International Carcinogenesis Meeting, held in Hanover, 22–24 October 1975, Edited by U. Mohr and others, International Agency for Research on Cancer, World Health Organisation (Geneva, 1977).

Damage to plants

The following books describe the effects of various air pollutants on plants:

The Effects of Air Pollution on Plants and Soil, Agricultural Research Council (London: HMSO, 1967). (Covers the effects of smoke, deposited matter, sulphur dioxide, fluorides, photochemical oxidants and lead)
Fume Damage to Forests, Forestry Commission Research and Development Paper No. 82 (1971). Forestry Commission, Alice Holt Lodge, Wrecclesham, Farnham, Surrey. (Summaries of 47 papers given at an international conference held in 1970)
Effects of Sulphur Oxides in the Atmosphere on Vegetation, revised chapter 5 of Air Quality Criteria for Sulfur Oxides. US Environmental Protection Agency (September 1973). EPA No. R3-73-030. (NTIS No. PB 226-314/AS)
Recognition of Air Pollution Injury to Vegetation, A Pictorial Atlas, edited by J. S. Jacobson and A. C. Hill. (Pittsburgh: Air Pollution Control Association, 1970)
Effects of Airborne Sulphur Compounds on Forests and Freshwaters, Pollution Paper No. 7. Department of the Environment, Central Unit on Environmental Pollution. (London: HMSO, 1976). (The British view of the effects of emissions of SO_2 in Scandinavia and the UK)
Air Pollution Across National Boundaries—The Impact of Sulphur in Air and Precipitation (and supporting studies), Sweden's Case Study for the UN Conference on the Human Environment. Royal Ministry for Foreign Affairs (Stockholm, 1972). (Sweden's account of the damage done in Sweden by SO_2 from the UK and elsewhere. The 'supporting studies' give the evidence in detail).

Guides to individual pollutants

If you need more detailed information than is provided by general reference books—for example, if you want to know the damage caused by a range of different concentrations of a pollutant—you can turn to more specialised reports dealing with individual pollutants. Most of these have been produced by various US government agencies.

AIR QUALITY CRITERIA REPORTS
A set of reports on six air pollutants was published in 1969 and 1970 by the US Department of Health, Education and Welfare (see table 2.1). Each report reviews in detail the main findings on the effects of the pollutant on man, animals, vegetation and materials; other useful chapters describe the physical and chemical properties of the pollutant, its sources and methods of measurement. These reports are particularly valuable because they contain concise summaries showing the known hazards of different concentrations of the pollutant.

This series of reports has been slightly extended and updated by the NATO-sponsored Committee on the Challenges of Modern Society. The NATO versions, issued between 1971 and 1974, contain more recent information than the American originals, but are rather less useful in that they lack indexes and summaries.

PRELIMINARY AIR POLLUTION SURVEY REPORTS
Evidence about the effects of a much wider range of pollutants is contained in another series of reports published in 1969 by the US Department of Health, Education and Welfare. These cover the same kinds of information as the Air Quality Criteria series, though in less detail, and include information on US and international air quality standards for each pollutant (see table 2.2).

BIBLIOGRAPHY AND ABSTRACTS REPORTS
This short series of reports published by the US Environmental Protection Agency contains references to, and summaries of, articles published on the effects of each pollutant on human health, plants, livestock and materials and includes a section covering 'standards and criteria' (see table 2.3).

GUIDES FOR SHORT-TERM EXPOSURES OF THE PUBLIC TO AIR POLLUTANTS
By 1976, the US Environmental Protection Agency had produced a short series of reports summarising the health effects and recommended exposure standards for each of three air pollutants (see table 2.4).

BIOLOGIC EFFECTS OF ATMOSPHERIC POLLUTANTS

The most detailed of all the reports in this section is a series produced by the US National Academy of Sciences. Each report runs to around 300 pages and covers the effects of one pollutant on man and the environment (see table 2.5).

CUEP POLLUTION PAPERS

Relatively few reports discussing the health and environmental effects of pollution are published in the UK. The main source of information is the Department of the Environment's Central Unit on Environmental Pollution—their reports on individual pollutants are shown in table 2.6.

TABLE 2.1
Air Quality Criteria reports, US Department of Health, Education and Welfare.[1] Available from NTIS (see page 172)

Title	Order number
Air Quality Criteria for Particulate Matter	PB–190–251
Air Quality Criteria for Sulphur Oxides	PB–190–252
Air Quality Criteria for Hydrocarbons	PB–190–489
Air Quality Criteria for Carbon Monoxide	PB–190–261
Air Quality Criteria for Photochemical oxidants	PB–190–262
Air Quality Criteria for Nitrogen Oxides	PB–197–333

Note
[1] Updated Air Quality Criteria reports with the same titles as those shown here (though combining the reports on photochemical oxidants and hydrocarbons) are available from: Committee on the Challenges of Modern Society, North Atlantic Treaty Organisation, Brussels 1110, Belgium.

TABLE 2.2
Preliminary Air Pollution Survey reports, US Department of Health, Education and Welfare. Available from NTIS (see page 172)

Pollutant	NTIS Order No.
Aeroallergens (pollens)	PB 188–076
Aldehydes (including acrolein and formaldehyde)	PB 188–081
Ammonia	PB 188–082
Arsenic and its compounds	PB 188–071
Asbestos	PB 188–080
Barium and its compounds	PB 188–083
Beryllium and its compounds	PB 188–078
Biological aerosols (microorganisms)	PB 188–084
Boron and its compounds	PB 188–085
Cadmium and its compounds	PB 188–086
Chlorine gas	PB 188–087
Chromium and its compounds (includes chromic acid)	PB 188–075
Ethylene	PB 188–069

Investigating Hazards

TABLE 2.2 Cont.

Hydrochloric acid	PB 188–067
Hydrogen sulphide	PB 188–068
Iron and its compounds	PB 188–088
Manganese and its compounds	PB 188–079
Mercury and its compounds	PB 188–074
Nickel and its compounds	PB 188–070
Odorous compounds	PB 188–089
Organic carcinogens	PB 188–090
Pesticides	PB 188–091
Phosphorus and its compounds	PB 188–073
Radioactive substances	PB 188–092
Selenium and its compounds	PB 188–077
Vanadium and its compounds	PB 188–093
Zinc and its compounds	PB 188–072

TABLE 2.3
(*Name of Pollutant*) *and Air Pollution. A Bibliography with Abstracts*, US Environmental Protection Agency. Available from NTIS (see page 172)

Title	NTIS order number
Odors and Air Pollution: A Bibliography with Abstracts 10/72	PB 312–733
Mercury and Air Pollution: A Bibliography with Abstracts 10/72	PB 214–011
Lead and Air Pollution: A Bibliography with Abstracts 1/74	EPA 4501–74–001
Nitrogen Oxides: An Annotated Bibliography 8/70	PB 194–429
Hydrocarbons and Air Pollution: An Annotated Bibliography (2 vols.) 10/70	PB 197–165
Asbestos and Air Pollution: An Annotated Bibliography 2/71	PB 198–394
Photochemical Oxidants and Air Pollution: An Annotated Bibliography (2 vols.) 3/71	PB 201–210
Chlorine and Air Pollution: An Annotated Bibliography 7/71	PB 203–355
Hydrochloric Acid and Air Pollution: An Annotated Bibliography 7/71	PB 203–341
Biological Aspects of Lead: An Annotated Bibliography (2 vols.)	PB 210–883 vol. 1 PB 210–884 vol. 2

TABLE 2.4
Guides for Short-term Exposures of the Public to Air Pollutants, US Environmental Protection Agency. Available from NTIS (see page 172).

Title	Order number
Basis for Establishing Guides for Short Term Exposures of the Public to Air Pollutants	PB–199–904
Vol. 1. Guide for Oxides of Nitrogen	PB–199–903
Vol. 2. Guide for Hydrogen Chloride	PB–203–464
Vol. 3. Guide for Gaseous Hydrogen Fluorides	PB–203–465

TABLE 2.5
Biologic Effects of Atmospheric Pollutants, US National Academy of Sciences. Available from: NAS, 2101 Constitution Avenue, Washington DC 20418, USA.

Title	Price (in 1976)
Lead	$7·75
Particulate Polycyclic Organic Matter	$7·75
Fluorides	$6·50
Asbestos	$3·75
Manganese	$6·25
Chromium	$6.50
Vanadium	$5·25
Nickel	$10·75
Vapor-Phase	$13·00
Selenium	$10·25

TABLE 2.6
Pollution Papers dealing with individual pollutants. Produced by the Central Unit on Environmental Pollution, Department of the Environment. Available from HMSO

	Title	Price
No. 2.	Lead in the Environment and its Significance to Man (1974)	45p
No. 5.	Chlorofluorocarbons and their Effect on Stratospheric Ozone (1976)	£1·00
No. 7.	Effects of Airborne Sulphur Compounds on Forests and Freshwaters (1976)	65p
No. 10.	Environmental Mercury and Man (1977)	£1·40

Sources of more recent information

BIBLIOGRAPHIES

You can find out whether any more recent reports on the effects of individual pollutants have been published in the US by consulting:

Air Pollution Technical Publications of the US Environmental Protection Agency (July 1974) plus subsequent quarterly bulletins. Available free of charge from EPA Air Pollution Technical Information Center. Research Triangle Park, North Carolina 27711, USA

Details of all Department of the Environment publications on pollution are shown in:

Sectional List 5. Department of the Environment, free from HMSO.

Investigating Hazards

A more comprehensively indexed guide to Department of the Environment publications during the last year is:

Annual List of Publications, free from Department of the Environment Library, 2 Marsham Street, London SW1.

Details of current research in the UK on the effects of environmental pollutants are given in the annual:

Register of Research. Part IV. Environmental Pollution, free from Headquarters Library (P3/008D) Department of the Environment, 2 Marsham Street, London SW1.

Current air pollution research is also shown each year in the:

National Society for Clean Air Year Book, NSCA, 136 North Street, Brighton. (Price of 1976 edition: £3.50)

AIR POLLUTION ABSTRACTS

If you have used one of the reference books or reports on individual pollutants listed in this section and want to know if any new information has been published since your main source appeared, you can consult a journal of abstracts. Abstracts journals summarise all recent reports on a particular substance—page 38 explains how to use them. Alternatively, and if you can afford it, you may want to use *TOXLINE*, the computerised up-to-date source of information on toxic hazards (see page 40).

Probably the best source of abstracts on air pollution subjects is *Air Pollution Abstracts* published every two months by the US Air Pollution Technical Information Center (do not confuse these abstracts with a less extensive series of abstracts, published, until a few years ago, under the same name by the UK's Warren Spring Laboratory).

Air Pollution Abstracts carries summaries of reports dealing with the effects of air pollutants on human health and the environment. It also covers many other aspects of air pollution including sources of emissions, control methods, techniques for measuring air pollution, economic aspects of air pollution, and standards and criteria.

Other journals which cover information about the health hazards of air pollution include:

Air Pollution Control Association Abstracts
Environment Index
Environmental Health and Pollution Control
Index to Air Pollution Research
Pollution Abstracts
UDS Air Quality Control Digest

Many of the abstracting journals dealing with the toxic hazards of chemicals used in industry, listed on page 39, will also include information on recent reports about the hazards of air pollutants.

If your pollutant is not mentioned, or not fully described, in the sources given here, look it up in one of the books on toxic hazards described in Part 5 of this chapter.

Evidence on the harmful effects of air pollutants to the community comes from the same kinds of studies as those used to predict the hazards faced by workers handling these substances. If you use a book that has been written with working hazards in mind, remember that the concentrations of a toxic substance in the air *outside* a factory are usually very much lower than those *inside* it. (Although external concentrations can become high—for example, if pollution control equipment breaks down or if pollution is trapped and unable to disperse because of atmospheric conditions—see page 79.)

The hazards to the outside community will therefore tend to be those caused by continued, long-term exposure (*chronic* effects) with less chance of finding the *acute* effects often caused by industrial exposure.

However, people living outside factories are likely to be more vulnerable than workers exposed to the same concentrations of a substance, because:

(1) workers are usually exposed to chemicals for perhaps eight hours at a time—the length of a normal working day. People living in an area suffering from air pollution are exposed continuously, 24 hours a day;

(2) workers are, on the whole, the fittest section of the community. People living outside include the very old, the very young and those who are too ill to work.

These differences explain why standards for exposure to air pollution are different from, and lower than, standards for chemical contamination of the factory air (see pages 87–8).

Part 9—Where to find information on water pollution hazards

Water pollution may affect various users of water in totally different ways. So before you start investigating the hazards of a water pollutant, decide who it is you think may be at risk.

If the risk is to people bathing in, or drinking water taken from, a polluted river:

(1) You will be interested in the effects of pollutants on man or, where this evidence is lacking, on animals used in laboratory testing. Use the sources of information in this section and, if necessary, those given in Part 5 of this chapter.

(2) When reading about the results of an experiment always check that the substance has been absorbed *orally*, whether by humans or by animals. This will give you some idea of the hazards of drinking or accidentally swallowing water polluted by the substance. Results from experiments in which the vapour of the substance is *inhaled* will not help you.

(3) Do not assume that a substance present in a river used as a source of drinking water will necessarily be present at the same concentration in drinking water itself. It may be removed by treatment at the water works or reduced in concentration by dilution with uncontaminated water.

If the risk is to fish or other organisms who may live in a polluted river:

(1) Information on the way human beings or mammals respond to the pollutant will not help you. You will need to find the results of studies in which fish or other aquatic organisms have been exposed to the pollutant.

(2) Hundreds of different kinds of species may live in the river, but studies on the effects of a particular pollutant may only have used one or two different species. Toxicity testing on fish normally involves the rainbow trout, a species thought to be more sensitive than other freshwater fish. Other species may be more resistant to the pollutant—though some may respond unexpectedly (see pages 137–8).

(3) The toxicity of a pollutant can be altered by many factors in the river which may not have been taken into account in laboratory experiments—some of these factors are described on pages 138–9.

(4) You may find that scientific publications use obscure Latin names when they refer to quite common fish or other river organisms; for example, by talking about *Gasterosteus aculeatus* when they mean 'three spined stickleback' or *Esox lucius* when they mean 'pike'. *The Penguin Dictionary of British Natural History* (now out of print but in many libraries) contains a straightforward index translating Latin names into common names, and other dictionaries and books on river life often provide a similar guide.

Several general reference books cover the hazards to man and to river life of a wide variety of different water pollutants; a number of guides dealing with individual pollutants contain more detailed information.

Reference books

Probably the single most useful book on the hazards of toxic substances to both drinking water and river life is:

Water Quality Criteria 1972, US Environmental Protection Agency (March, 1973). Ecological Research Series. EPA No. R3-73-033. $12.80.

The drinking water hazards of 43 different pollutants are described in section 2 of this book, which includes wherever possible:

(a) the normal concentration of each pollutant found in lakes, rivers and drinking water supplies in the US;
(b) details of the hazards reported to have been caused by drinking water polluted by particular substances;
(c) a summary of the hazards known to be associated with each toxic substance and a description of the symptoms of acute and chronic poisoning in man;
(d) an estimate of the total normal human intake of each substance from various sources and a description of its fate in the body;
(e) a recommended drinking water standard for each substance.

The toxicity of various pollutants to freshwater life is described in section 3 of the book which includes details of (for each substance if possible):

(a) the concentrations of the substance found in the tissues of fish or other aquatic species;
(b) the toxicity of the substance to different forms of aquatic life;
(c) a recommended water quality standard.

Two extracts from *Water Quality Criteria 1972* are shown in figures 2.10 and 2.11. Other sections of this book deal specifically with the effects of pesticides and water pollution hazards to marine life.

An enormous amount of scientific evidence on water pollution hazards is summarised in:

Water Quality Criteria Data Book, US Environmental Protection Agency Water Pollution Research Series.
Vol. 1: Organic Chemical Pollution of Freshwater EPA No. 18010DPV 12/70 (1970)
Vol. 2: Inorganic Chemical Pollution of Freshwater EPA No. 18010DPV 07/71 (1971)

A Summary of Some Effects of pH on Freshwater Fish and Other Aquatic Organisms

pH	Known effects
11.5-12.0	Some caddis flies (Trichoptera) survive but emergence reduced.
11.0-11.5	Rapidly lethal to all species of fish.
10.5-11.0	Rapidly lethal to salmonids. The upper limit is lethal to carp (Cyprinus carpio), goldfish (Carassius auratus), and pike. Lethal to some stoneflies (Plecoptera) and dragonflies (Odonata). Caddis fly emergence reduced.
10.0-10.5	Withstood by salmonids for short periods but eventually lethal. Exceeds tolerance of bluegills (Lepomis macrochirus) and probably goldfish. Some typical stoneflies and mayflies (Ephemera) survive with reduced emergence.
9.5-10.0	Lethal to salmonids over a prolonged period of time and no viable fishery for coldwater species. Reduces populations of warmwater fish and may be harmful to development stages. Causes reduced emergence of some stoneflies.
9.0-9.5	Likely to be harmful to salmonids and perch (Perca) if present for a considerable length of time and no viable fishery for coldwater species. Reduced populations of warmwater fish. Carp avoid these levels.
8.5-9.0	Approaches tolerance limit of some salmonids, whitefish (Coregonus), catfish (Ictaluridae), and perch. Avoided by goldfish. No apparent effects on invertebrates.
8.0-8.5	Motility of carp sperm reduced. Partial mortality of burbot (Lota lota) eggs.
7.0-8.0	Full fish production. No known harmful effects on adult or immature fish, but 7.0 is near low limit for Gammarus reproduction and perhaps for some other crustaceans.
6.5-7.0	Not lethal to fish unless heavy metals or cyanides that are more toxic at low pH are present. Generally full fish production, but for fathead minnow (Pimephales promelas), frequency of spawning and number of eggs are somewhat reduced. Invertebrates except crustaceans relatively normal, including common occurrence of mollusks. Microorganisms, algae and higher plants essentially normal.
6.0-6.5	Unlikely to be toxic to fish unless free carbon dioxide is present in excess of 100 ppm. Good aquatic populations with varied species can exist with some exceptions. Reproduction of Gammarus and Daphnia prevented, perhaps other crustaceans. Aquatic plants and microorganisms relatively normal except fungi frequent.
5.5-6.0	Eastern brook trout (Salvelinus fontinalis) survive at over pH 5.5. Rainbow trout (Salmo gairdneri) do not occur. In natural situations, small populations of relatively few species of fish can be found. Growth rate of carp reduced. Spawning of fathead minnow significantly reduced. Mollusks rare.
5.0-5.5	Very restricted fish populations but not lethal to any fish species unless CO_2 is high (over 25 ppm), or water contains iron salts. May be lethal to eggs and larvae of sensitive fish species. Prevents spawning of fathead minnow. Benthic invertebrates moderately diverse, with certain black flies (Simuliidae), mayflies (Ephemerella), stoneflies, and midges (Chironomidae) present in numbers. Lethal to other invertebrates such as the mayfly. Bacterial species diversity decreased; yeasts and sulfur and iron bacteria (Thiobacillus-Ferrobacillus) common. Algae reasonably diverse and higher plants will grow.
4.5-5.0	No viable fishery can be maintained. Likely to be lethal to eggs and fry of salmonids. A salmonid population could not reproduce. Harmful, but not necessarily lethal to carp. Adult brown trout (Salmo trutta) can survive in peat waters. Benthic fauna restricted, mayflies reduced. Lethal to several typical stoneflies. Inhibits emergence of certain caddis fly, stonefly, and midge larvae. Diatoms are dominant algae.
4.0-4.5	Fish populations limited; only a few species survive. Perch, some coarse fish, and pike can acclimate to this pH, but only pike reproduce. Lethal to fathead minnow. Some caddis flies and dragonflies found in such habitats; certain midges dominant. Flora restricted.
3.5-4.0	Lethal to salmonids and bluegills. Limit of tolerance of pumkinseed (Lepomis gibbosus), perch, pike, and some coarse fish. All flora and fauna severely restricted in number of species. Cattail (Typha) is only common higher plant.
3.0-3.5	Unlikely that any fish can survive for more than a few hours. A few kinds of invertebrates such as certain midges and alderflies, and a few species of algae may be found at this pH range and lower.

FIGURE 2.10
Extract from *Water Quality Criteria 1972*

Sublethal doses of inorganic chemicals for aquatic organisms

Constituent	Chronic dose	Species	Conditions	Literature citation
Arsenic (As) (See also Sodium (Na) and Potassium (K))				
	20 ppm	Salmo gairdneri and minnows	conc. of arsenic using sodium arsenite, fish overturned in 36 hrs.	Grindley 1946[188]
	250 ppm	....	conc. of arsenic using sodium arsenate, fish overturned in 16 hrs.	"
	30-35 ppm	minnows	fins, scales damaged, diarrhea, heavy breathing and hemorrhage around fin areas.	Boschetti and McLoughlin 1957[187]
	4-10 µg	Mytilus edules	amount of As retained in flesh	Sautet et al. 1964[230]
	0.5-2 µg	Mytilus edules	amount of As retained in shell when exposed to 100 g/l of As.	"

FIGURE 2.11
Extract from *Water Quality Criteria 1972*

Vol. 3: Effects of Chemicals on Aquatic Life EPA No. 18050GWV 05/71 (1971)

Vol. 4: An Investigation into Recreational Water Quality EPA No. 18040DAZ 04/72 (1972)

Vol. 5: Effects of Chemicals on Aquatic Life. (An updated version of Vol. 3) EPA No. 18050HLA 09/73 (1973).

Nearly all the published literature on water pollution hazards available at the time of publication is summarised in these volumes. However, unlike *Water Quality Criteria 1972*, the *Water Quality Criteria Data Book* presents nearly all of its information in the form of tables without providing any comment or interpretation, and makes no attempt to recommend limits for substances in rivers or drinking water.

The toxic effects to man and laboratory animals of substances absorbed orally are described in volumes 1 and 2. Tables show:

(a) typical concentrations of substances in freshwater and their sources;

(b) concentrations reported in human and animal tissue;

(c) a summary of the acute and chronic toxicity of the substances;

(d) the maximum concentration of each pollutant thought to produce no effect in man after long-term exposure;

(e) lists of substances thought to produce cancer, mutations or birth defects that have been detected in water.

Volume 2 also contains an extremely valuable guide to the limitations of the evidence collected and warnings against misinterpretation.

Investigating Hazards 55

The effects of chemicals, especially pesticides, on fish and aquatic life are summarised in volume 3 and updated in volume 5.

A rather older book, which tends to interpret the data it presents rather more than the Data Book, is *Water Quality Criteria*.[5] A short summary of the hazards of 90 different substances present as water pollutants or air pollutants is given in *Pollution Detection and Monitoring Handbook* (described on page 43). A very brief coded summary attaching a toxicity rating to over 500 substances to aquatic life is contained in Appendix 8 of *Registry of Toxic Effects of Chemical Substances* (described on page 28). Equally brief information on 300 different substances is given in a guide to dealing with accidental releases of hazardous materials issued by the US Coastguard.[6] This guide also briefly summarises the fire, explosion and health hazards to man of the substances it lists and advises on emergency procedures to deal with any discharge.

Guides to individual pollutants

The hazards of individual pollutants are dealt with in detail by several series of specialised reports.

EIFAC REPORTS
A set of reports published by the European Inland Fisheries Advisory Commission (EIFAC), a body sponsored by the UN's Food and Agricultural Organisation, deal with the effects of eleven substances on freshwater fish. The reports review the evidence on the accumulation of the substances studied in fish tissue, their lethal and sublethal effects on fish and their impact on other freshwater organisms. The reports conclude by recommending 'water quality criteria' which, if observed, should give complete protection to freshwater fish (see page 136). The titles of EIFAC reports published by 1977 are shown in table 2.7.

WATER POLLUTANT BIBLIOGRAPHIES
A series of reports containing references to the effects of various pollutants in water has been published by the US Water Resources Scientific Information Center (see table 2.8).

NOTES ON WATER RESEARCH
The UK's Water Research Centre produces series of leaflets covering many aspects of water pollution, including the toxicity of individual pollutants to fish. A list of these notes (formerly called 'Notes on Water Pollution') and reprints of papers published by Water Research Centre staff can be obtained from: The Director, Water Research Centre,

Stevenage Laboratory, Elder Way, Stevenage, Herts (telephone: 0438 2444).

TABLE 2.7

Titles of *EIFAC reports on pollution and freshwater fish*. Available from Publications Division, Food and Agricultural Organisation of the United Nations, Via delle Terme di Caracalla, 00100 Rome, Italy

Title, EIFAC Technical Report Number and date
Report on finely divided solids and inland fisheries (*1*) (*1964*)
Report on extreme pH values and inland fisheries (*4*) (1968)
Report on water temperature and inland fisheries based mainly on Slavonic literature (*6*) (1968)
List of literature on the effect of water temperature on fish (*8*) (1969)
Report on ammonia and inland fisheries (*11*) (1970)
Report on monohydric phenols and inland fisheries (*15*) (1972)
Report on dissolved oxygen and inland fisheries (*19*) (1973)
Report on chlorine and freshwater fish (*20*) (1973)
Report on zinc and freshwater fish (*21*) (1973)
Report on copper and freshwater fish (*27*) (1976)
Report on cadmium and freshwater fish (*30*) (1977)

TABLE 2.8

(*Name of Pollutant*) *in Water. A Bibliography*, US Water Resources Scientific Information Center (Port Washington, New York)

Name of pollutant	WRSIC number
Aldrin and Endrin	72–203
Arsenic and Lead	71–209
Cadmium	73–209
Chromium	72–205
Copper	71–204
DDT	71–211
Detergents	71–214
Dieldrin	72–202
Magnesium	71–206
Manganese	71–205
Mercury	72–207
PCB	72–201
Strontium	71–201
Trace elements	71–202
Zinc	71–208

INFORMATION SERVICE ON TOXICITY AND BIODEGRADABILITY (INSTAB)

The Water Research Centre provides an information service covering the effects of more than 1500 different chemicals on sewage treatment processes and fish and other aquatic life.

Investigating Hazards 57

The INSTAB service exists mainly to help manufacturers and water authorities locate information about the hazards of effluent discharges; they may be willing to deal with requests for information from members of the public who have unsuccessfully tried to obtain information about water pollution hazards from other sources. If they do agree to help with a request they will normally supply a list of references and abstracts of published papers on the hazards of a pollutant. INSTAB can be contacted at the Water Research Centre, Stevenage Laboratory, Elder Way, Stevenage, Herts (telephone: 0438 2444).

Sources of more recent information

Details of recent Department of the Environment publications on pollution and of current research in the UK on the effects of water pollution are shown in the sources listed on pages 48–9. References to occasional US government publications on water pollution hazards are shown in the regular *ORD Publications Summary* available from the Environmental Protection Agency's Office of Research and Development, Cincinnati, Ohio 45268.

ABSTRACTS JOURNALS
References to recent papers in scientific journals on the hazards of water pollutants can be found either by using a journal of abstracts (details of how to use these journals are given on page 38) or by consulting *TOXLINE*, the computerised source of information on toxic hazards (see page 40).

The best journal for abstracts on water pollution is probably the Water Research Centre's *WRC Information* (formerly known as Water Pollution Abstracts). This thoroughly indexed journal deals with water pollution hazards and all other aspects of water pollution. Other abstracting journals which may contain information on the toxicity of water pollutants include:

Environment Index
Environmental Health and Pollution Control
Health Aspects of Pesticides Abstracts Bulletin
Pesticides Review
Pollution Abstracts
Robert A. Taft Water Research Centre. Selected Summaries of Water Research
Sport Fishery Abstracts
UDS Water Quality Control Digest
World Fisheries Control Abstracts

Some of the toxic hazards abstracting journals shown on page 39 may also cover recent information on the hazards of water pollutants.

If your pollutant is not adequately covered by the sources described in this section, try using the books listed in Part 5 of this chapter. Remember that only studies dealing with toxic substances absorbed *orally* are directly relevant to water pollution hazards.

Air Pollution

The air pollution section of this Handbook is divided into three chapters.

Air Pollution Law and Standards (chapter 3) explains the legal control of industrial air pollution in the UK and the policy of using emission limits and tall factory chimneys to keep pollution at ground level to acceptably low concentrations. Chapter 3 tells you how to find out (a) what standards apply to a particular factory, (b) whether the factory's emissions comply with the legal limits, and (c) whether the resulting ground level pollution is likely to exceed the predicted 'acceptable' concentrations under certain conditions.

Air Quality Objectives (chapter 4) describes the informal 'objectives' used to help decide whether the concentrations of pollution in the air are satisfactory. It reviews the various different objectives and suggests how you can use them yourself and find out how much protection they are likely to give.

Air Pollution Monitoring (chapter 5) tells you how to find out how much pollution is in the air around you. If the right kind of pollution monitoring has been done, chapter 5 will help you use the results to discover (a) how much a particular factory is adding to air pollution levels, (b) whether existing concentrations of pollution are likely to damage human health or the environment (sources of information on the toxicity of air pollutants are given in Part 8 of **Investigating Hazards**—chapter 2), and (c) whether pollution control policies have succeeded in reducing air pollution.

3 Air Pollution Law and Standards

A law dating back over 1,000 years in Guernsey has been invoked by Mr. William McAlister. He went on his knees and raised the 'Clameur de Haro' by crying: 'Haro, Haro, Haro, come to my aid, O Prince, I am being done a wrong', and reciting the Lord's Prayer in French. He did this to stop a rose-growing firm. . . . from operating a heating boiler said to be depositing soot on his premises and affecting his family's health. The plea was registered at the registry office at St. Peter Port and within a year the case has to be heard by the island's Royal Court.

Daily Telegraph (25 August 1976)

Summary

The control of industrial air pollution in the UK is divided between the central government Alkali Inspectorate, which deals with air pollution from certain manufacturing processes, and local authorities, which deal with air pollution from other industrial sources.

Factories using processes controlled by the Alkali Inspectorate are required, under the Health and Safety at Work Act, to use the 'best practicable means' (a) to prevent emissions, and (b) to ensure that any emissions that do escape do not cause a hazard or a nuisance.

Outline 'best practicable means' are published for each industry but the Alkali Inspectorate does not disclose either the final details of the standard imposed on individual factories or information about emission levels. Information about emissions from *some* of these factories will be available from local authorities, as they are able—but not obliged—to publish this data.

Air pollution from furnaces is dealt with by local authorities under the Clean Air Acts, which limit emissions of smoke, grit and dust. Local authorities have powers to deal with other pollution only if it can be shown to have caused a hazard or a nuisance.

The standards enforced by local authorities are all on public record, but emissions data will be published only if local authorities choose to release it.

Both the Alkali Inspectorate and local authorities control the heights of chimneys used to discharge pollution so that by the time they reach the ground emissions are diluted to concentrations believed to be relatively safe. The calculations used to set chimney heights are not exact or valid under all weather conditions; at times ground level pollution may be much greater than predicted and, occasionally, hazardous.

Air quality versus emission standards

There are two alternative systems for controlling air pollution. The first is to set standards showing how much pollution can be safely tolerated in the air and to ensure that no factory—or other source—is allowed to discharge so much pollution that these 'air quality standards' are exceeded. This roughly corresponds to the system used to control river pollution in the UK. It is not used in air pollution control because there are said to be too many difficulties both in setting 'safe' standards and in predicting how emissions from individual sources will affect air quality.[1,2] The use of air quality standards is described in chapter 4.

The alternative approach, which is used in the UK, is to set fixed standards of emission which limit the concentration of pollutants discharged and, in some cases, to require factories to use the best pollution control methods available at a reasonable cost. As new pollution control equipment is developed, firms may be required to use it—even if the air around the factory does not seem to be heavily polluted. This system of control focuses on the standard of emissions and not on the quality of the air, though special action is taken to prevent any emissions that might cause a health hazard.

Pollution control authorities

Air pollution control is divided between central and local government. **The Alkali Inspectorate,** part of the national Health and Safety Executive, deals with pollution from a large number of industrial processes (the 'scheduled processes') including many of the most troublesome sources of air pollution such as lead, iron and steel and other metal industries, petroleum refineries, electricity and gas works, cement works and a wide range of chemical processes. A company operating any scheduled process is required to register it with the Alkali Inspectorate; these factories are known as *registered works*.

Air pollution from any unregistered works is dealt with by **local authorities'** Environmental Health Departments, under different legislation. Environmental Health Officers control smoke, grit and dust pollution from furnaces and any other form of industrial air pollution that causes a hazard or nuisance.

Start your enquiries into factory pollution by finding out what kind of processes go on at the factory and which pollution control body is responsible for enforcing standards.

Often, a company operates both registered and unregistered processes at the same site, so control over the factory's air pollution emissions is divided between the Alkali Inspectorate and the local authority. Ask the company or the local authority whether any registered processes are involved. A full list of the 60 or so kinds of registrable processes is available[3] and this may help you decide whether a particular process comes under Alkali Inspectorate control.

The Alkali Inspectorate and the registered works

Some aspects of the Alkali Inspectorate's system of control were being reconsidered by the government at the time of going to press (March 1978). The Alkali Inspectorate was transferred from the Department of the Environment to the Health and Safety Executive in 1974, and much of the 1906 Alkali Act—under which the Inspectorate formerly operated—had been replaced by the Health and Safety at Work Act. In 1976, the Royal Commission on Environmental Pollution recommended that the Alkali Inspectorate be returned to its former home in the Department of the Environment, though with somewhat different powers and policies.[4] The system described here is that which was in force in March 1978.

Best practicable means (bpm)

There is no set of fixed legal standards for emissions from registered works. Instead, anyone operating a registered works is required by the Health and Safety at Work Act to:

> use the best practicable means for preventing the emission into the atmosphere from the premises of noxious or offensive substances and for rendering harmless and inoffensive such substances as may be so emitted (*section 5(1)*).

According to this definition, the 'best practicable means' (bpm) have to be used to do two things. First of all, to prevent emissions, whether they come from chimneys or from any other source in the factory. And, secondly, to ensure that any pollutants that are emitted do not cause a hazard or a nuisance.

The term 'best practicable means' is not fully defined in the Act. However, the HSWA states that 'means': 'includes a reference to the manner in which the plant provided for those purposes is used and to the supervision of any operation involving the emission of the (noxious or offensive) substances' *(section 5(2))*.

Thus, firms are not simply required to use the 'best practicable' pollution control equipment but must also make sure that their staff is properly trained to operate it and that precautions to prevent emissions are used to deal with, for example, stockpiles of dusty raw materials, or contaminated air extracted from inside the factory.

'Practicable' is defined in the 1974 Control of Pollution Act, in the context of controlling noise, as: 'having regard among other things to local conditions and circumstances, to the current state of technical knowledge and to the financial implications' *(section 72(2))*.

This definition is not legally binding when it comes to air pollution from registered works, but it gives a fair idea of the factors the Alkali Inspectorate traditionally considers when setting standards. Hence, the cost of pollution controls are taken into account, and the Inspectorate has stated that 'the technically possible would be impracticable if the costs were so high that the manufacturing operation were thereby rendered unprofitable or nearly so.'[5]

Guidelines explaining the components of 'best practicable means' for each industry have in the past been decided by the Chief Alkali Inspector, although there may be much wider public consultation in the future.[6] Traditionally, these industry-wide guidelines have been made up of four parts:

(1) An emission limit setting the maximum permitted concentration of dust or gas that may be released from the chimney. These standards have no legal force but the Alkali Inspectorate normally assumes that if they are met, then the 'best practicable means' of controlling chimney emissions have been used. Failure to meet these standards has not necessarily been taken to mean that the bpm have not been used. The Royal Commission on Environmental Pollution has reported of a case where:

> a certain registered works emitted dark smoke for a considerable period in contravention of the agreed bpm. The emission was witnessed by two Environmental Health Officers who were prepared to give evidence in a prosecution. But because it could not be demonstrated what had caused the emission of dark smoke and thus that bpm had not been used, the prosecution could not be pursued.[7]

(2) The required sampling frequency. Registered works are required, as part of the 'best practicable means', to monitor the concentration of chimney discharges. Where adequate automatic monitoring equipment

has been developed, firms must use it to give a continuous picture of emissions. Otherwise, the guidelines will state how often samples must be taken. The frequency varies according to the industry; it may be as often as once a day, or so rarely (for example, once a year for some electricity works[8]) as to give very little useful information.

(3) Guidance on setting the chimney height. The bpm have to be used to ensure that any pollution that is emitted is 'harmless and inoffensive'. This is done by discharging gases and dusts from chimneys tall enough to dilute harmful substances to 'safe' levels by the time they reach the ground. The guidelines will often give a formula or state the principles to be used in calculating chimney heights.

(4) General instructions. Registered works do not meet their legal obligations simply by keeping emissions from chimneys within the Inspectorate's limits. They also have to use the best practicable methods of controlling emissions from other parts of the factory even if emission limits cannot be set. The Inspectorate's guidelines will lay down the general principles to be followed in minimising pollution from these sources.

Note that under this system firms are only required to use certain methods to try to prevent pollution, but do not have to avoid pollution at any cost. **Indeed, registered works often cause considerable nuisance even though they are using what the Alkali Inspectorate considers to be the 'best practicable means'.**

Two things can be done when the 'best practicable means' fail to protect the community from pollution. If emissions are likely to cause 'a risk of serious personal injury' the Alkali Inspectorate can shut down the process by issuing a 'prohibition notice' (Health and Safety at Work Act, *section 22*). But if the pollution only causes a nuisance—and not a health hazard—the Alkali Inspectorate is often unwilling to demand that firms meet stricter standards. In such a case the local authority can apply to the Secretary of State for the Environment for permission to take nuisance action against the registered works. If a successful action is brought in the High Court, the firm may be ordered to prevent the nuisance or else close down.

Sources of information about 'best practicable means'

Notes on the 'best practicable means' for different industries have in the past been published in the old annual reports of the Chief Alkali Inspector. A summary of emission limits in force in 1975, and a guide showing where to find full 'best practicable means' for each industry, was published in 1977.[9] In future, 'best practicable means will be issued as codes of practice under the Health and Safety at Work Act.

The published guidelines for 'best practicable means' in each industry do not contain full details of the standards applied to individual factories. These may vary from the published guidelines, because:

(1) Registered works are subject to the standards in force when they were built, or when their equipment was last replaced, and not to the most recent standards. New 'best practicable means' are published from time to time, but these apply only to newly built works, which are not permitted to begin production until they can meet these standards.[10] Existing works do not have to adopt the new standards until their current pollution control equipment comes to the end of its useful life and is due for replacement or major overhaul.

(2) To take local conditions into account, the actual standard applied to a works will often differ slightly from the industry-wide guidelines. This variation will usually not apply to the emission limit, but will affect the height of the chimney and possibly some other factors.

(3) Factories using processes recently brought under Alkali Inspectorate control are sometimes given several years' breathing space before the full standards are applied. Sometimes this deadline is extended for individual works. In other cases, where there are 'persistent justified complaints' about the factory's pollution, the Inspectorate states that it will require the new standard to be adopted immediately. However, this has not always been done.[11]

TABLE 3.1
Some national emission limits for cadmium in 1974[12]

Country	Standard* (mg/m^3)
Australia	3
New South Wales, Australia	20
Japan	1
Singapore	20
United Kingdom†	39

Notes

* In some cases the standard shown applies to any source of cadmium emissions and in other cases only to certain industrial processes.

† The UK standard includes a maximum limit on the weight of cadmium that may be discharged during a week (13·6 kg per 168-hour week).

The details of 'best practicable means' applied to an individual registered works is decided by the District Alkali Inspector after consulting the firm (there may be wider public consultation in future).

So do not assume that the published national guidelines are followed in all ways by a particular factory without checking with the company or the Alkali Inspector.

You may sometimes want to see how the Alkali Inspectorate's emission standards compare to those used in other countries. A full list of emissions standards used throughout the world in 1974 has been published in an American book called *The World's Air Quality Management Standards*[12] (see table 3.1).

Sources of information about emissions

The Alkali Inspectorate is well known for its unwillingness to release information about pollution from individual registered works. It has justified this policy at various times by arguing that information about emissions will be meaningless to the lay person, disclose trade secrets to competitors, be deliberately abused by 'extremists in the environmental movement' and that greater disclosure will harm the trust that exists between the Inspectorate and industry. This secretiveness has been strongly criticised (see, for example, *Social Audit's* report on the Alkali Inspectorate[11]). The Royal Commission on Environmental Pollution called the Inspectorate's policy on information 'misguided'. While it strongly supported many aspects of the Inspectorate's work it concluded that much of the public criticisms of the Inspectorate:

> stem from real deficiencies in the system, and that to a considerable extent they have their origins in the fact that the Inspectorate have not sufficiently adapted to changes in society's attitude to pollution and to public accountability . . . they sometimes appear remote and autocratic. There has been some clumsiness and insensitivity in the Inspectorate's public pronouncements and an air of irritation with those who presume to question the rightness of their decisions.[13]

Unfortunately, the 1974 Health and Safety at Work Act has given legal backing to the Inspectorate's policy of confidentiality. The Act prohibits Inspectors from giving information obtained from registered works to the public unless it has the firm's permission to do so.[14] The Royal Commission on Environmental Pollution has called for this legal restriction to be lifted.

If you want to know how much pollution a registered works is producing, do not ask an Alkali Inspector. Local authorities, on the other hand, will be able to help you.

Under the Control of Pollution Act, a local authority has the right to obtain and publish information about air pollution emissions from all works in its area—those controlled by the

Alkali Inspectorate as well as those that come under the authority's own control.

The Act does not *require* local authorities to collect this information—but it gives them the powers to obtain it if they decide they want it.[15] Once they use these powers to obtain information they must publish full details on a public register which you will be able to inspect during normal working hours.[16]

Ask your local authority to use these powers. It can ask for information about most pollutants produced by a non-registered works, though a registered works can only be asked to supply the same information it normally gives to the Alkali Inspectorate.[17]

The local authority can ask for the following information from a registered or non-registered works:

(1) the total length of time in any given period for which a pollutant was emitted;

(2) the temperature, speed and rate at which gases are emitted from a chimney or outlet, and the height of the discharge;

(3) the total amount of the pollutant discharged during a specified period and the average concentration. The concentration of sulphur dioxide produced by combustion does not have to be supplied as this can be estimated from the sulphur content of the fuel.[16]

The local authority can also obtain details of past emissions, providing records have been kept.

Not all works will have to supply all the above information to local authorities. If the information is not 'immediately available' or cannot be compiled without unreasonable expense a firm can appeal to the Secretary of State for the Environment, and, if it is successful the firm will not have to supply the information requested by the local authority.[18] This exemption will not usually affect Alkali Inspectorate-controlled works, since the local authority can only ask for information that is already being collected and supplied to the Inspectorate. The exemptions may mean that in many cases local authorities will only be able to obtain results of occasional 'spot checks' on emissions.

Companies may also appeal against having to disclose information if they feel that to do so would (a) harm their commercial interests by revealing trade secrets (and this is possible only in a tiny minority of cases), or (b) be contrary to the public interest.[19] The 'public interest' exemption will probably apply only to defence and military establishments.[20]

Special local committees may be set up by local authorities before they start publishing information about emissions from individual works. The Department of the Environment has recommended that these committees should represent the local authority, local industry and 'local persons with suitable qualifications to offer advice on atmospheric pollution issues'.[21] Their comments may be published in the public registers, alongside the emissions data.

In future, there may be another source of information on air pollution from both registered and unregistered works. Regulations can be made under the Health and Safety at Work Act (*section 3(3)*) requiring firms under certain circumstances to publish specified information about the effects of their activities on the health or safety of the public.

Local authorities and the non-registered works

Air pollution from any industrial process not registered with the Alkali Inspectorate is dealt with by local authorities.

The most common source of pollution from non-registered works is from furnaces used to heat boilers. Most factories use a boiler to heat the premises, to provide heat for a manufacturing process, or to generate electricity. Emissions of smoke or solids from a boiler are controlled under the 1956 and 1968 Clean Air Acts. Any other form of air pollution that may cause a nuisance or a health hazard is dealt with under the 1936 Public Health Act. Local authorities may also attach conditions limiting air pollution to planning permission for a new factory.

Sources of information on standards

You can judge whether a non-registered works has met its legal obligations to control air pollution by referring to three sets of standards:

(1) Has the company obtained the 'prior approval' of the local authority for its plans to build or operate a new process and has it met any relevant planning conditions?

(2) Have emissions from any furnace complied with the fixed legal standards laid down in the Clean Air Acts and Regulations?

(3) Have emissions caused a nuisance or health hazard to the local community?

Prior approval

Before a firm installs a new furnace or chimney, it is required, under the Clean Air Acts, to notify the local authority and satisfy it that (a) the furnace can operate smokelessly,[22] (b) if necessary the furnace is fitted with equipment to remove grit and dust from emissions,[23] and (c) the chimney used to

carry away the emissions is tall enough to prevent a nuisance or health hazard.[24]

The local authority's Environmental Health Officer (EHO) may be prepared to discuss with you how well the company's pollution controls are working and you may be able to see the official forms requesting and granting permission for the proposed furnace and chimney. (These will, for example, allow you to calculate the rate at which SO_2 is being emitted.) Since the Control of Pollution Act, local authorities have been permitted to disclose information of this sort providing that no trade secrets are involved.[25] Start by making sure that the company has actually obtained the necessary approval for any new furnace or chimney:

> In 1974, Avon Medicals Ltd., a Birmingham firm, was found to have installed a new boiler without first notifying the local authority, as was required by law. Birmingham City Council only learned of this after an enquiry from *Social Audit*, when it was discovered that the firm had erected a chimney of only 28 feet height: a 40-foot chimney should have been built.[26]

Find out from the EHO or from the council's planning department whether conditions referring to air pollution or to nuisance were imposed when planning permission was granted. Conditions of this sort sometimes require factories to meet certain emission standards or to monitor concentrations of air pollution around the factory.

At the same time, ask whether a public planning enquiry was held before planning permission was given. If it was, the evidence to the enquiry and the report of the planning inspector may contain useful guides to the amount of pollution the factory was expected to create. The company may have given assurances that air pollution would be restricted in certain ways, and these assurances may now be a useful yardstick with which to judge the factory's current performance.

Planning consent and conditions can be obtained from the local authority. The report of any public inquiry should also be available from the local authority, though it may be more difficult to locate copies of evidence given at the enquiry. Try the Department of the Environment, the company itself, a local newspaper, or anyone else giving evidence.

Smoke, grit and dust standards

No factory is allowed to emit dark smoke, except for specified short periods.[27] If the factory is located in a smoke control area (you can find out where these are from the local authority) it is not allowed to produce *any* smoke at all, even light smoke.[28]

The Clean Air Regulations allow certain exemptions to these prohibitions: they are explained in detail in the *National Society for Clean Air Year Book*[29] and in *Clean Air—Law and Practice* by Garner and Crow.[30]

If you think a chimney is emitting dark smoke for more than a few minutes a day, keep a note of the length of time you observe it. Your Environmental Health Officer will have to confirm that the smoke is actually 'dark'. He may judge this from experience or by comparing it to a series of progressively darker cards known as the 'Ringelmann Chart' or on an instrument based on these cards.[31, 32]

Maximum limits have been set for grit and dust emissions from furnaces if the pollution is caused by the fuel being burned and not by the material being heated.[33] Other furnaces must use 'any practicable means' to prevent emissions,[34] though fixed standards for incinerators, cupolas and other furnaces have been recommended by an official working party and can be used as guidelines.[35] Standards for fume emissions may also be laid down in the future.[36]

The terms 'grit', 'dust' and 'fume' all refer to solid particles of different sizes. 'Grit' is defined as particles of more than 76 microns diameter (a micron is a millionth of a metre), anything between 1 and 76 microns is classed as 'dust', while solid particles smaller than 1 micron are described as 'fume'.

Emissions of sulphur dioxide are controlled mainly by dispersing them from tall chimneys, designed to dilute the gas to low concentrations by the time it reaches the ground. The method of calculating chimney heights is described later in this chapter.

Sulphur dioxide emissions can be regulated by controlling the amount of sulphur present in fuel. A legal limit has been imposed on the sulphur content of gas oil—one kind of industrial fuel oil—in order to comply with an EEC directive.[37] From the beginning of 1977 no furnace has been permitted to burn gas oil containing more than 0·8 per cent sulphur, and this limit will be reduced to 0·5 per cent from October 1980. The limits apply to registered works as well as to non-registered works—but power stations are not included. The limits are not expected to have any immediate impact on sulphur dioxide pollution since gas oil used in Britain at the time these regulations were introduced was, on average, already within the first stage limit.

Smells or other nuisance

Other forms of pollution from non-registered works can be dealt with by local authorities if they cause a health hazard or a nuisance. The 1936 Public Health Act requires local authorities to inspect their areas for nuisances and to take action to prevent them.

This is a significant change of emphasis from the approach of other air pollution legislation which ignores the impact of emissions on the community, so long as the factory meets certain emission standards or uses approved pollution control methods.

The local authority can instruct a factory to take action to end a nuisance or to prevent it recurring.[38] If it still continues, the authority can apply to the Magistrate's Court for a 'nuisance order' requiring the firm to end the nuisance; a private individual can also apply for a nuisance order in this way.[39] However, the firm may not have to end the nuisance if it can show that it has already used the 'best practicable means' available. Where the nuisance is caused by unpleasant smells a report on the 'best present practice' in controlling industrial odours may be referred to as a guide to 'best practicable means'.[40]

If the local authority believes that 'best practicable means' are not enough to prevent the nuisance it can take action against the firm in the High Court—where the use of 'best practicable means' is not a defence.[41]

The penalties for failing to comply with a High Court order to end a nuisance are much more severe than the penalties that may be imposed in a Magistrate's Court. The owner of a factory that continues to cause a nuisance in defiance of a High Court order may be gaoled for contempt of court, and in several cases factories have been shut down altogether.[42]

Although a local authority normally has no powers to interfere in the control of a registered works dealt with by the Alkali Inspectorate, it can sometimes take nuisance action—providing it first obtains the consent of the Secretary of State for the Environment. In this way, registered works could be forced either to meet much stricter standards than those normally imposed by the Alkali Inspectorate or to close down.

Sources of information about emissions

You can find out how much pollution is being emitted from the chimney of a non-registered works from your local authority—provided it has used its powers to obtain and publish this information.

The Control of Pollution Act allows a local authority to collect information about air pollution from non-registered works either by (a) going into the factory and taking its own samples, or (b) requiring the firm to monitor emissions and supply it with the results.[43]

The kinds of details that can be asked for, the procedure for publishing the information, and the cases in which firms can withhold information are described on page 68.

Information about potential pollution

If your local authority has not published anything about air pollution from a particular factory you can get some general idea of the kinds of pollution likely to be involved from the Chief Alkali Inspector's annual reports[44] and various textbooks on air pollution.

Many American sources publish *emission factors*.[45] These tell you the weight of pollutant that would be discharged, if no pollution control equipment was used, when a given quantity of fuel is burned or a certain weight of product is manufactured.

> *Example:* for every 1000 gallons of fuel oil containing 3 per cent sulphur the uncontrolled emission would typically contain approximately 450 lbs of sulphur dioxide, 100 lbs of nitrogen oxides, 8 lbs of solids and 5000 µg of the carcinogen benzo(α)pyrene.[46]

These emission factors are very approximate and refer to American fuels and production methods. You may sometimes find them useful as a guide to kinds and quantities of pollution that might be emitted elsewhere. Remember that if pollution control equipment is being used, much of the potential pollution will be prevented.

In many cases, pollution control equipment will remove *some* of the pollutants produced by a process. A description of how pollution control equipment works is given in most air pollution textbooks or can be obtained from the companies that manufacture the equipment. Manufacturers will often provide useful publicity material on their products and may also be prepared to answer specific questions about their equipment. Names and addresses of the manufacturers of 'gas cleaning equipment' are given in the *National Society for Clean Air Year Book*.[29]

Some of the modern equipment used to remove dusts and solids from chimney gases is extremely efficient and in some cases can remove up to 99·9 per cent of solid particles.

The key points to remember about very efficient pollution control equipment are that:

(1) The tiny percentage of pollutants that escape may still represent an enormous amount. A large cement or steel works may release several tons of dust into the air every day even using the best dust control equipment.

(2) A slight change in the efficiency of the equipment can have a drastic effect on the amount of pollution discharged. A drop in efficiency from 99·9 per cent to, say, 99·6 per cent actually means that four times more pollution is being allowed to escape.

(i.e. 0·4 per cent of the pollution reaching the device is passing through it instead of only 0·1 per cent.)

Try and find out from the company itself what kind of pollution control equipment it is using, which pollutants in the emissions it removes and which are left untouched, what efficiency it was designed to produce and what its present efficiency is. The manufacturers of this and similar equipment may then be able to tell you what this equipment should be able to achieve at its best and what could be done (and at what cost) using the most efficient devices available.

The most widely used emission factors describe the quantity of sulphur dioxide produced when different fuels are burned. Table 3.2 shows the weight of sulphur dioxide (SO_2) produced by burning different fuels of the kind used in the UK to give the same amount of heat.

In the UK, SO_2 is not removed from chimney gases before they are discharged: all the SO_2 produced when the fuel is burned goes into the air. To calculate the amount of SO_2 given off you must know (a) the average sulphur content of the fuel (normally stated as a percentage), (b) the proportion of sulphur normally given off when the fuel is burned, and (c) the rate at which the furnace consumes fuel.

Example
How much SO_2 is emitted in an hour from a furnace burning 10 tons/h of coal containing 1·5 per cent sulphur?

The weight of sulphur consumed in an hour is

$$\frac{10 \times 1 \cdot 5}{100} = 0.15 \text{ tons}$$

If only 90 per cent of this sulphur is emitted (see table 3.2) this becomes

$$\frac{0 \cdot 15 \times 90}{100} = 0 \cdot 135 \text{ tons}$$

Sulphur is converted to sulphur dioxide. Every atom of sulphur (molecular weight 32) combines with two atoms of oxygen (molecular weight 16). A molecule of SO_2 therefore weighs $\frac{64}{32}$ times more than an atom of sulphur. Hence the weight of SO_2 given off when 10 tons/h of coal is burned is

$$\frac{0 \cdot 135 \times 64}{32} = 0 \cdot 27 \text{ tons}$$

TABLE 3.2
Comparison of the emissions of sulphur dioxide when coal and oil are burned in different appliances[47]

Application	Fuel	Gross calorific value Btu/lb	Sulphur content per cent	Proportion of sulphur liberated as SO_2 per cent	Efficiency of appliance per cent	Production of an amount of useful heat equal to that given by 1 tonne of coal of 12 800 Btu/lb burnt at 100 per cent efficiency			
						Fuel required tonne	SO_2 liberated, lbs		NO_2, lbs
							Solid fuel	Oil	
Domestic heating									
open fire, old-fashioned type	coal	12 800	1.6	80	25	4.00	228	—	—
open fire, improved type	coke	12 000	1.2–1.4	80	40	2.67	114–134	—	3–8
closed stove	coke	12 000	1.2–1.4	80	65	1.64	72–83	—	—
	anthracite	14 500	1.0	80	65	1.36	49	—	—
boiler (including domestic central-heating boiler)	coke	12 000	1.2–1.4	80	60	1.78	76–90	—	10
	anthracite	14 500	1.0	80	75	1.18	43	—	—
flueless heater	vaporising oil	19 600	0.06–0.16	100	75	0.87	—	2.0–6.3	3–10
	kerosine	20 000	0.03–0.04	100	90	0.71	—	0.9–1.3	
Central heating									
solid fuel boiler	coal	12 000	1.6	90	75	1.42	92	—	—
	coke	12 000	1.2–1.4	90	75	1.42	69–81	—	10
	anthracite	14 500	1.0	90	78	1.10	45	—	—
oil-fired boiler, less than 500 000 Btu/h	oil	19 600	0.75–0.90	100	75	0.87	—	29–36	3–10
oil-fired boiler, greater than 500 000 Btu/h	oil	18 600	1.4–2.8	100	80	0.86	—	54–108	
Steam-raising (industrial)	coal	12 000	1.6	90	75	1.42	92	—	10–30
	coke	12 000	1.2–1.4	90	75	1.42	69–81	—	
	oil	18 400	2.5–3.8	100	80	0.87	—	96–150	10–30
Electricity generation	coal	10 200	1.6	90	86	1.46	94	—	10
	oil	18 300	2.0–3.0	100	86	0.81	—	110–150	20

Chimneys and pollution

The use of tall chimneys to dilute pollutants is a key part of the way both the Alkali Inspectorate and local authorities control pollution. Releasing gases or dusts high above the ground reduces the amount that comes down in any one place. The principle which is often quoted is 'there are no such things as harmful substances, only harmful concentrations'. Pollution authorities hope that by calculating the right height for the chimney they can keep ground concentrations of pollution to within certain levels for most of the time.

The 'tall chimney policy' is sometimes alleged to be a method of avoiding having to use expensive pollution control equipment to prevent emissions. When a tall chimney is presented as an answer to pollution problems people often wonder whether the complex calculations they are based on are really accurate, whether the 'safe' concentrations they are expected to produce are really safe, and whether sending dusts or gases out through tall chimneys only spreads the problem farther afield instead of removing it.

This section describes the reasoning behind the use of tall chimneys in pollution control. It does not try to show you how to carry out your own chimney height calculations, but it does point out the assumptions made by those people who do. And it shows you how to think about, and where necessary question, the value of these calculations.

In normal weather conditions, gases or very fine dusts released from the tops of chimneys are mixed with enormous volumes of air before they reach the ground. Two powerful factors, the wind and upward-moving currents of air, help to dilute the emissions.

To picture the effect of *the wind* on pollution, think of a thin stream of smoke coming from the ground in a wind of 4·4 metres/second (m/s) (10 mph). This trickle of smoke will have spread out into a cloud 36 m (40 yards) wide by the time it has travelled 91 m (100 yards) from its source. Had it been released above the ground, it would also have spread out above and below its point of discharge.

Chimney gases are best dispersed in moderate winds: the 'plume' from the chimney touches down relatively far away from its source and the maximum concentration of pollution is relatively low. In strong winds the plume may be swept down to the ground quite close to the chimney, while in very light winds or calm conditions the gases may sink to the ground around the stack. A study of sulphur dioxide concentrations near Tilbury power station found that the maximum concentrations of the gas occurred when the wind speed was either more than 10 m/s (22 mph) or less than 4 m/s (8·9 mph).[48]

Predictions of the effect of chimney emissions on ground concentrations of pollution always assume that the wind will blow at a certain speed or range of speeds. If you are looking at one of these predictions, find out what wind speed has been used—and then compare it to records of actual wind speeds in the area (these are produced by the Meteorological Office). **When the actual speed of the wind is greater or lower than the values used in the calculation, the ground level concentration of pollution may be much higher than predicted.**

Even when there is no wind, *warm air currents* will carry away and dilute pollution. During a warm day the sun heats up the ground which in turn warms the air near it. This hot air rises until it cools to the temperature of the surrounding air. On a typical summer's day warm air currents can travel up to several thousand metres, greatly diluting the pollution they carry with them.

Sometimes this natural mixing of the air is blocked by what is known as a *temperature inversion*. Normally, the warmest air is near the ground and the air above it is progressively cooler; in an inversion this process is reversed and a layer of cool air is trapped underneath warm air. Inversions are quite common in some regions, especially at night. If pollution is caught underneath an inversion the normal dispersion and dilution of emissions is prevented. **None of the normal methods used to predict the impact of chimney emissions on ground concentrations of pollution apply during conditions of inversion: the actual concentrations of pollution may greatly exceed those predicted.**

The effect of the chimney

The taller the chimney is, the more time its emissions have to disperse before they reach the ground. Any slight increase in the height of the chimney will produce a proportionally much greater reduction in ground level concentrations of pollution: this concentration is proportional to the square of the height from which they are released.

In theory, doubling the chimney height will cut ground level concentrations of pollution not to a half but to a quarter; trebling the chimney height will cut pollution on the ground to a ninth.

Anything that increases the height that the gases reach after leaving the chimney will cause a proportionally greater decrease in ground level pollution. Chimney gases are often hot, and therefore continue to rise for some time after leaving the chimney; the speed of the gases when they leave the chimney (the *exit velocity*) also helps. The final height reached by the plume after leaving the chimney is known as the *effective chimney height*.

Any increase in the effective chimney height has two effects: (1) it decreases the resulting ground level concentration, and (2) it moves the point at which the maximum concentration occurs further away from the chimney.

Two rough and very approximate rules-of-thumb, which apply only under conditions of moderate wind speeds and normal dispersion, are:

(1) the maximum ground level concentration of pollution caused by a chimney emission occurs at a distance of somewhere around 15 times the effective chimney height.[49]

(2) the maximum concentration of pollutant, averaged over half an hour to an hour, is given by

$$C_{max} = \frac{23\,000 \times R}{H^2}$$

where the maximum concentration is in $\mu g/m^3$; R is the rate at which the pollutant is emitted, in g/s; and H is the effective chimney height in metres.

A formula for calculating the 'effective chimney height' is given in 'Gaseous Pollution from Chimneys'.[50]

The height of the chimney becomes absolutely crucial during times of inversion. If an inversion forms near the ground, anything discharged underneath it—for example, pollution from domestic chimneys—will be trapped. Sometimes chimney gases may be carried for several miles under an inversion barrier until they suddenly reach higher ground, or are brought down, when they produce massive concentrations.

Taller chimneys, attached to factories, may be able to discharge

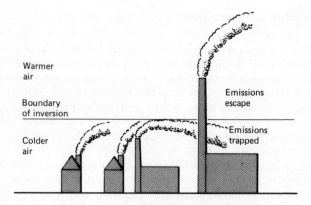

FIGURE 3.1
Low-level domestic and industrial emissions trapped under an inversion while high-level emissions escape[51]

their gases *above* the inversion layer (see figure 3.1). Alternatively, gases from a chimney below an inversion may have enough 'lift' to force their way through the layer.

Unfortunately, some inversions may reach up to several hundred metres above the ground—while others begin quite high up and reach even further. Find out from the Meteorological Office how common inversions at different heights are for your area. If the inversions occur or extend above the effective chimney height of a factory or power station you can expect emissions to be trapped underneath them; concentrations may then become much greater than predicted and in extreme cases may become dangerous. A temperature inversion was partly responsible for the London 'smog' of 1952 when 4000 people died during and shortly after several days of extremely high concentrations of air pollution.

If the chimney gases are said to be able to force their way through an inversion because of their 'lift', check to see what gas temperature and velocity must be maintained for this to happen. Then ask under what conditions, and how frequently, the temperature and velocity may be lower than is necessary (for example, when the plant is starting up or not running at full capacity). This leads to other questions: does the company monitor the temperature and velocity of the escaping gases? Does it know when inversions occur? Does it have any plans to reduce emissions at times when the inversion layer is higher than normal or when the chimney gases are unable to penetrate the inversion?

Where does the pollution end up?

At first sight, the most likely effect of using tall chimneys to disperse emissions is simply to reduce local concentrations of pollution at the expense of higher concentrations elsewhere.

The results of sulphur dioxide monitoring in the UK between 1961 and 1971 tend not to support this view.[52] The amount of SO_2 discharged during the period increased because of a growth in power station emissions; much of this SO_2 was discharged from extremely tall chimneys. Yet despite the increase in high level emissions, the measured concentrations of SO_2 at ground level dropped by nearly 30 per cent. This decrease seems to mirror the reduction in domestic and low level emissions over the period, suggesting either that high level emissions do not have much effect on pollution at ground level or that emissions are coming down in unmonitored areas.

About half the sulphur dioxide emitted by British industry is thought to be absorbed into the ground or washed out in the rain over the UK. Most of the rest is probably washed out in rain falling into the sea. However, a significant quantity of the gas is carried over long distances, converted to sulphuric acid in the atmosphere and deposited in the form of acidic rain over Scandinavia.

Sweden and Norway believe that an increase in the acidity of rain caused by air pollution from Britain and elsewhere may have contributed to a reduction in forest growth and to the decline of freshwater salmon and trout in their countries.[53, 54, 55] To find out what really does happen to sulphur dioxide from tall chimneys, large scale studies on the transport of SO_2 in the atmosphere are now being carried out.[56]

Predicting ground level pollution

The effect of chimney emissions on pollution concentrations at any given point can be predicted by mathematical formulae which try to estimate the influence of many different varying factors.[57] These include the direction and speed of the wind, the amount of mixing that takes place in the atmosphere both in a horizontal and vertical direction, the temperature and speed of the chimney gases, the effects of nearby buildings and the surface of the land, and any physical and chemical changes that affect the pollutants. Not surprisingly, this calculation involves very complicated mathematics and a thorough knowledge of local atmospheric and other conditions.[58, 59]

A rough method of estimating the average monthly concentration of pollution caused by chimney emissions at any distance from the chimney has been developed.[50]

Researchers at Warren Spring Laboratory have also been developing methods of predicting the combined effect over a wide area of many different sources of air pollution, though it is unlikely that these will turn out to be simple to apply.[60]

Chimney heights for registered works are set by the Alkali Inspectorate, which uses a mixture of mathematical formulae and its own experience of the behaviour of chimney plumes under different conditions.[61]

Chimney heights for non-registered works are determined by local authorities who use a set of simple charts published by the Department of the Environment.[62] These apply only to furnaces burning coal or oil but not natural gas; nor do they apply when pollutants more toxic or more likely to cause nuisance than sulphur dioxide are emitted.

Chimney heights calculated according to these charts are adjusted to take into account existing levels of pollution in the neighbourhood. The charts call for taller chimneys to be built in more polluted areas, so that a new factory will add slightly less to ground level pollution than a similar factory elsewhere. Table 3.3 shows the maximum concentration of pollution, as a three-minute average, that a chimney conforming to the recommended heights should add to the air. These figures cannot be directly compared to existing air pollution monitoring results which normally show only the 24-hour average concentration. The three-minute figure can very roughly be translated into the equivalent 24-hour concentration by dividing it by ten—provided that weather conditions are 'average'.[64] Hence, under normal weather conditions these chimneys would add a maximum of between 40 and 60 $\mu g/m^3$ of sulphur dioxide to existing concentrations. These calculations assume that exit velocity of the chimney gases is maintained above a certain level; if they are not, the 'lift' of the emissions is reduced and ground level pollution may be greater than predicted.

TABLE 3.3

Maximum additional concentrations of SO_2 at ground level expected, under normal weather conditions, from chimneys conforming to the *Chimney Heights Memorandum*[63]

Category of district*	Maximum additional concentration of SO_2, as a three-minute average ($\mu g/m^3$)
(A) Unpolluted	570
(B) Low pollution	500
(C) Moderate pollution	460
(D) Considerable pollution	430
(E) Severe pollution	400

Note
* Full definitions of the types of district represented by categories (A) to (E) are given in the *Chimney Heights Memorandum*.[62]

How accurate are predictions?

None of the formulas or charts used to predict ground level concentrations of pollution is very accurate. They are normally said to be within a factor of two of the actual concentration; this means that the real concentration of pollution may be as much as double the predicted value or as little as one-half of it. The Central Electricity Generating Board has measured SO_2 concentrations around some of its power stations and reported that a simple method of predicting concentrations 'produces a figure which is not exceeded in practice for more than one or two per cent of the time and is never exceeded by more than a factor of two'.[65]

At other times the difficulties in accurately estimating all the different meteorological factors have led to very large errors. A study which compared the predicted and measured concentrations of SO_2 around a sulphuric acid plant in Germany, found that after taking more than

2000 measurements over a two-year period 'the calculations yield significantly smaller concentrations than the measurements'. The prediction was that 95 per cent of all measurements would be below 30 µg/m³, but in fact the 95 per cent level was around 300 µg/m³—ten times higher. The average of all the measurements taken downwind of the stack turned out to be 50 times greater than the predicted average.[66]

How to question predictions

At public planning enquiries, and at other times, you may have the opportunity to discuss the likely impact of air pollution from a new factory or process. At these occasions, the company or the pollution control authority will usually predict the likely impact of the factory's emissions on ground level pollution.

You are unlikely to have the expertise to check that these calculations have been carried out properly or to produce your own, alternative, calculations. However, if you understand the thinking behind the predictions you should be able to question them effectively.

The following list of questions may help you look for the extra details that will tell you how accurate and relevant the predictions that have been made really are.

(1) How wide is the area that will be affected by the new emissions and does it contain groups of people (in hospitals or schools, for example) who may be particularly vulnerable to any additional air pollution?

(2) Are the existing concentrations of pollutant in this area already being monitored and what concentrations have been found? The average and the maximum concentration should always be given, and it may be useful to know the maximum concentration found in 95 per cent of all measurements.

(3) How do existing concentrations of pollution in the area compare to those known to be hazardous to man or damaging to the environment?

(4) Is existing monitoring adequate (see chapter 5) and can it be relied on to give a fair picture of ground level pollution? If not, are there good reasons to believe that existing pollution concentrations are low and can be added to safely?

(5) How much will emissions from the new chimney add to existing concentrations of pollution? The forecasts should show the concentrations likely to be found at varying distances and directions from the chimney, especially in populated areas and in areas where pollution concentrations are already high. The length of time that the highest

concentrations are likely to persist should be given, and values for the average, maximum and 95 per cent concentrations should be shown. Different averaging times (three minutes, daily, seasonal, annual etc.) should be shown so the results can be easily compared to monitoring data from other sources.

(6) How do the expected concentrations of pollution compare to levels known to be hazardous to man or dangerous to the environment?

(7) What extra height of chimney would be needed to reduce the estimated ground level concentration by any given amount, and what would be the cost of such an extension? How much would it cost to prevent these emissions reaching the air by installing pollution control equipment?

(8) Not all factory pollution is emitted from chimneys. What is the likely contribution of emissions from other sources to ground level pollution?

(9) How accurate are the predictions of additional ground level pollution from the chimney? If they are accurate to within a factor of two, for example, try doubling all the predicted concentrations and seeing how they then compare to concentrations known to be hazardous.

(10) Under what conditions will the predictions not apply (such as very strong winds, periods of calm, inversions, emissions at low temperature and velocity)?

(11) How often will the conditions mentioned in question 10 above be likely to occur?

(12) When the normal predictions do not apply, what will be the maximum possible ground level concentrations of pollution at different points, and what likely effects will they have on human health or the environment?

(13) Will the company be able to detect conditions under which the predictions are invalid as soon as they occur, and does it have any special plans for reducing emissions at these times?

(14) Will special monitoring of ground level concentrations of pollution around the new chimney be carried out in order to confirm that the predicted concentrations of pollution are not exceeded, and will the results of such monitoring be made public?

(15) If, after the chimney has been built, the impact of its emissions is found to be much greater than predicted, will action be taken to reduce emissions?

4 Air Quality Objectives

Summary

Although there are no legally enforceable standards for air quality in the UK, informal 'air quality objectives' are sometimes used. These include a set of objectives recommended by an Expert Committee of the World Health Organisation and the 'acceptable' ground level concentrations of pollutants that chimney heights are calculated to produce.

Enforceable air quality standards are used in some countries and may be adopted in the UK if current EEC proposals are accepted. However, the Royal Commission on Environmental Pollution has recommended the use of a flexible system of air quality guidelines as an alternative to the EEC proposals.

Whether or not they form part of a formal system of air pollution control, air quality objectives can be used as a quick and valuable guide to 'acceptable' concentrations of pollution. However, they should not be used without first checking what, if any, degree of nuisance or risk they may incorporate.

Setting air quality standards

Air pollution control in the UK is based on emission standards —limits on the amount of pollution that can be discharged by individual factories. There is relatively little emphasis on the quality of the outside air and no legally enforceable 'air quality standards'. In some cases the concentrations of certain factory pollutants in the air outside the factory are measured: but unless pollution causes a very serious health hazard the factory is not normally required to reduce its emissions if their impact has been greater than expected.

'Our approach is to attack emission sources, from which an acceptable general air quality follows' argued the Department of the Environment in a paper published in 1977.[1] The system assumes that if factories use the best pollution control methods they can reasonably afford, the quality of the air will normally remain within reasonable limits.

At present, there is no mechanism for assessing the extent of air pollution, deciding when concentrations are too high, and requiring emissions from industry or other sources to be reduced. Ideally, this means knowing at what points concentrations of pollution first become harmful, though this would require more information than is normally available (see chapter 1).

Trying to set air quality standards invariably leads to controversy. Industry demands firm proof that existing concentrations are harmful—the public may feel that emissions should be prevented unless industry can show that existing levels of air pollution are safe.

Both sides turn to the scientists for an independent view—and find that they are unable to give absolute answers. An EEC committee which in 1975 considered a proposed air quality standard for lead reported:

> It is difficult to set legally binding standards on a scientific basis . . . scientists could not say precisely which levels would represent a hazard . . . Moreover, almost nothing is known of the complex effects of many harmful substances in combination and it . . . could not be proved that certain limits should be respected.[2]

Instead of trying to set legally enforceable standards, pollution authorities may refer to 'air quality criteria' and 'air quality objectives':

(1) *Air quality criteria* are simply pieces of scientific evidence about the relationship between pollution concentrations and damage. They have no legal or administrative significance. Criteria would, for example, tell you at what concentrations pollution damage to plants, animals and human health had been detected.

(2) *Air quality objectives* are concentrations of pollution in the air which governments, international bodies, or pollution control agencies recommend should not be exceeded in order to give a certain level of protection against particular pollutants. (They are also sometimes known as 'goals' or 'guidelines'.) No-one will usually claim that any objective (other than zero pollution) will give *complete* protection to everyone. Some 'acceptably low' risk is generally involved.

This chapter describes the air quality objectives sometimes used in the UK and abroad. Use this chapter with chapter 5, which tells you how to find out how much pollution is in the air, and with chapter 1, which explains how the evidence about health hazards is collected and why there is no such thing as a 'safe' standard. **Try not to use any air quality standard or objective unless you know how much protection—and what degree of risk—it assumes is acceptable.**

WHO goals

A set of air quality objectives, produced by an expert committee of the World Health Organisation in 1972, is often used in the UK.[3]

The WHO report contains air quality criteria—evidence about the hazards of different concentrations—for five pollutants: sulphur oxides, smoke, carbon monoxide, photochemical oxidants and nitrogen dioxide (the last three of these are more commonly emitted by motor vehicles than by industry). The report invites countries to use these criteria in

TABLE 4.1
Air quality criteria and goals for sulphur dioxide (SO_2) and smoke[3]

Daily average concentration	Effects
500 $\mu g/m^3$ SO_2 with 500 $\mu g/m^3$ smoke[a]	Increased deaths amongst people suffering from heart or respiratory disease
500 $\mu g/m^3$ SO_2 with 250 $\mu g/m^3$ smoke[a]	Worsening in condition of people suffering from bronchitis
250 $\mu g/m^3$ SO_2 with 250 $\mu g/m^3$ smoke[a]	Increased symptoms in people suffering from respiratory disease
200 $\mu g/m^3$ of sulphur oxides with 120 $\mu g/m^3$ of smoke should not occur on the same day for more than seven to eight days during the year [a, b, c]	World Health Organisation Expert Committee's recommended long-term goal
Annual average concentration	*Effects*
100 $\mu g/m^3$ SO_2 with 100 $\mu g/m^3$ smoke[a]	Increased frequency of respiratory disease and symptoms in children
86 $\mu g/m^3$ SO_2	Chronic plant injury and excessive leaf drop[d]
80 $\mu g/m^3$ smoke	Complaints of annoyance and reduced visibility
60 $\mu g/m^3$ of sulphur oxides with 40 $\mu g/m^3$ smoke[a, c]	World Health Organisation Expert Committee's recommended long-term goal

Notes
[a] Values for smoke and SO_2 apply only in conjunction with each other.
[b] The permitted seven to eight days may not include any consecutive days.
[c] Although the Expert Committee's criteria refer to SO_2 the goals are actually stated in terms of sulphur oxides.
[d] Source: *Air Quality Criteria for Sulphur Oxides*, US Department of Health, Education and Welfare (January 1969).

setting 'short-term goals', in other words, targets to aim for immediately. The report does recommend long-term goals—actual concentrations which countries should adopt as standards or objectives at some time in the future when they can afford them.

The long-term goals represent concentrations which the committee believed 'would be unlikely to produce any ill-effects at all . . . in the light of present knowledge'. They covered four of the five pollutants (nitrogen dioxide was excluded) dealt with in the earlier part of the report.

Some of these goals are commonly used in the UK as informal guidelines. The Greater London Council has incorporated three of them into its guidelines for London's air quality,[4] adding objectives for ozone and particulate lead. Warren Spring Laboratory has produced contour maps of the UK showing which areas in 1972–73 suffered from smoke or sulphur dioxide levels above the WHO long-term goals.[5, 6] A summary of the air quality criteria and WHO goals for smoke and SO_2 is shown in table 4.1.

Chimney height objectives

Chapter 3 explained how local authorities and the Alkali Inspectorate control the heights of chimneys used to disperse factory emissions. Chimney heights are calculated to dilute emissions so that concentrations of pollution on the ground remain within certain levels for most of the time.

The amounts of sulphur dioxide that a new non-registered works, with a coal or oil-fired boiler, is permitted to add to existing concentrations is shown in table 3.3 on page 81. However, the ground level concentrations of sulphur dioxide which this calculation assumes are acceptable are not explicitly stated.

Chimney heights calculated by the Alkali Inspectorate for registered works aim to produce ground level concentrations of pollution which do not exceed one-fortieth of the Threshold Limit Value (TLV)—the concentration to which workers inside factories may be exposed (see pages 12–15). The TLV/40 figure is explained in two stages:

(1) the TLV is divided by four because people living near a factory are exposed to pollution continuously throughout the week, not just for the 40-hour working week on which TLVs are based;

(2) this figure is then divided again, by ten, to allow for the very old, young or ill people in the community who are more vulnerable to pollution than the supposedly healthy adult workers for whom TLVs are intended.

The TLV/40 figure is often quoted as a 'guideline' for community exposure, although those who use it stress that is an arbitrary figure 'which should not be considered as a recommended limit.'[7]

The *drawbacks* of this approach are that:

(1) *It assumes that the TLVs provide adequate protection for workers.* As is shown in chapter 1, they sometimes do not.

(2) *TLVs are designed to prevent damage to human health, whereas in some cases nuisance and damage to the environment occur at concentrations which do not harm human beings.* The US government has recognised this by adopting two sets of air quality standards: one designed to prevent damage to human health and a stricter second set designed to protect the environment.

(3) *The Alkali Inspectorate publishes very little information about its standards.* In some cases the Inspectorate has said that the TLV is divided by more or less than 40; however, the public is not told which air quality objective the Inspectorate has chosen on its behalf.

(4) *Pollution concentrations around factories are not usually monitored to see that they conform to the TLV/40 figure.* Even when they are monitored, firms are not normally required to reduce emissions when the measured concentrations exceed the air quality objective.

Air quality guideline bands

The need to 'focus attention openly and specifically on air quality' in the UK system of air pollution control was recognised by the Royal Commission on Environmental Pollution in its fifth report.[8]

The Commission considered whether a system of legally enforceable air quality standards, similar to that used in the United States, would be helpful and decided that this would be costly, complicated and imprecise. Expensive new monitoring programmes would be needed to check on air quality, it might not be possible to prove where excess pollution was coming from, and in many cases arbitrary decisions would be made on how several sources should share out the reduction in emissions needed to achieve the air quality standard.

Instead, the Commission recommended a more informal system of *air quality guideline bands* which would not be legally enforced but would serve as goals and as an aid to planning.

A national guideline band—showing a range of concentrations and not just a single figure—would be established for various air pollutants. The upper level would represent the 'highest tolerable' concentration, and emissions would normally have to be reduced if this figure was exceeded. The lower level of the band would represent a wholly satisfactory air quality 'below which there would generally be no justification for pressing for a reduction in emissions'.

Within this wide national guideline, local authorities would choose their own narrower *target bands*, depending on the extent of the local problem and other local factors (see figure 4.1).

The target levels would generally fall within the (national) guideline band; the long-term aim would no doubt be to get comfortably below the upper guideline level but there would often be great practical difficulty and little justification for attempting to reach the lower level.[9]

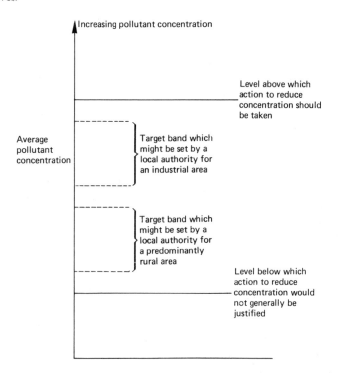

FIGURE 4.1
A national air quality guideline band and local target bands as proposed by the Royal Commission on Environmental Pollution

The proposed main use of the new guidelines and targets would be in controlling new developments. Local authorities would use them in deciding whether planning permission for a new factory should be refused because it would lead to an unacceptable increase in air pollution or whether new housing should be forbidden in high pollution areas because of the risk to the health of the inhabitants. The new targets would also, it was hoped, allow local authorities to plan the reduction of high pollution levels. However, under the present pollution control system, factories can only be required to reduce emissions if they cause a hazard or nuisance—but not if they infringe an informal target level set by the local authority.

Elsewhere in its report, the Royal Commission suggested changes in air pollution law that would allow pollution authorities greater powers to vary their standards and to tighten them up if ground level pollution was too high. All industrial discharges—including those from factories currently dealt with by local authorities—would be required to use the 'best practicable means' to prevent pollution, and details of the standard applied to each factory would be kept on a publicly available 'consent document'. The consent would be revised every two or three years, and if necessary stricter conditions could be written in.[10]

At the time of writing (March 1978) the government had not decided whether to accept the Royal Commission's proposals.

EEC standards

The European Economic Communities (EEC) is committed, under its environmental programme, to a series of steps in which it first publishes air quality criteria for air pollutants, then agrees on 'quality objectives' and finally introduces air quality standards which are to be legally enforced throughout the EEC countries.[11]

The European Commission has already published draft standards for lead,[12] and for smoke and sulphur dioxide,[13] and has plans to draft standards for a variety of other air pollutants, including nitrogen oxides, carbon monoxide, photochemical oxidants, asbestos, hydrocarbons and vanadium.

If all EEC countries agree to these standards, they will be required to put them into practice by a fixed deadline and report to the Commission on their progress towards meeting them.

The EEC proposals were discussed by the Royal Commission on Environmental Pollution in its fifth report. The Commission concluded that they would 'be unenforceable in practice and would bring the law into disrepute'. Since then, the Department of the Environment has stated that the EEC's subsequent proposals have tended to emphasise 'objectives' rather than 'standards' and that 'in the Department's view the concept as now envisaged by the EEC is close to the Royal Commission's concept of guidelines'.[14] The latest status of the EEC proposals can be learnt from the Commission's Press and Information Office.[15]

US standards

Legally enforceable air quality standards are in use in the United States where, in 1971, national standards for six pollutants (particulates, sulphur oxides, carbon monoxide, nitrogen dioxide, photochemical oxidants and hydrocarbons) were set. Two sets of standards are used: primary standards, designed to safeguard human health, and secondary standards, which aim to prevent damage to materials and the environment. The secondary standards are either equal to or stricter than the primary standards.

Each state was required to draw up plans showing how emissions would be reduced in order to bring air pollution down to the legal limits. In 1976, most areas were expected to meet the primary standards by the early 1980s, though it was estimated that in 1975 68 million Americans were exposed to air that exceeded the primary air quality standard for particulates and 13 million people lived in areas where the sulphur oxides standard was exceeded.[16]

Planners are advised not to allow new residential development in areas where pollution exceeds a primary standard by a certain factor[17] and, originally, no new source of pollution was allowed to come into operation if air pollution in the area exceeded one of the relevant national standards. This policy has since been modified:

> The (US) Environmental Protection Agency has announced a new 'trade off' policy under which industrial development will be permitted in areas where the air has not yet been cleaned up to the level required by law. The announcement ... brought immediate protests from environmental groups. ...
>
> The policy would allow a steel mill, for example, to build a new plant in an industrial area if the steel company—or another company—reduced pollution from another plant in the same area. ...
>
> (An official) said the new approach was a 'compromise' between dictates of the clean air act, which prohibits the introduction of new sources of pollution such as power plants in areas where air quality standards have not been met, and the demands for industrial development in some of those same areas.[18]

Details of US air quality standards can be found in most American air pollution reference books or textbooks published after 1971.

International standards

Air quality standards are also used in the USSR, where the aim is to define concentrations at which no direct or indirect harmful or unpleasant effects are caused in man including those that would normally be imperceptible or be considered harmless by authorities in Europe and the United States. A list of 114 'maximum permissible concentrations of harmful substances in the air of population centres' has been published.[19]

However, scientists in the USSR acknowledge that their health protection standards 'may, of course, be above or below that attainable in practice and ordinarily found in the air. For this reason, as a temporary expedient, sanitary or technological standards (which take into account technical and economic feasibility) may be used as a guideline.'[20]

Many other countries have adopted air quality standards or objectives. A complete list of those in use during 1974 is given in *The*

TABLE 4.2
Air quality standards for lead in 1974

Country	Standard ($\mu g/m^3$)	Averaging time	Comment
Bulgaria, Czechoslovakia, USSR[a]	0·7	24 hours	Includes compounds of lead other than tetraethyl lead
Canada (Ontario)[a]	15·0	24 hours	
Israel[a]	5·0	24 hours	
Italy[a]	10·0	8 hours	
Italy[a]	50·0	30 minutes	Not to be exceeded more than once in eight hours
USA (California)[b]	1·5	30 days	
USA (Montana)[b]	5·0	30 days	
Greater London Council[c]	3·0	3 months	For lead in suspension
EEC proposed[d]	2·0	annual mean	For urban residential and areas exposed to sources of lead other than motor vehicle traffic
EEC proposed[d]	8·0	monthly median[f]	For areas particularly exposed to motor vehicle traffic
TLV for factory air in UK[e]	150·0	8 hours	For exposure during a 40-hour working week

Notes
[a] Source: *The World's Air Quality Management Standards, Volume 1*.[21]
[b] Source: *The World's Air Quality Management Standards, Volume 2*.[21]
[c] Source: *Greater London Intelligence Quarterly*.[4]
[d] Proposal for a Council Directive on air quality standards for lead[12].
[e] Source: Threshold Limit Values for 1976.[22]
[f] The median is the concentration below which half of all samples taken fall.

Air Quality Objectives 93

World's Air Quality Management Standards[21] published by the US Environmental Protection Agency (see table 4.2).

Using air quality objectives

Once you have found out how much pollution is in the air (chapter 5 will help you do this) you will want to know if this concentration is likely to be a hazard to human health or to damage any part of the environment. The most reliable way to do this is to look the pollutant up in the sources given in chapter 2; but referring to an accepted air quality objective may save you time and effort. To do this:

(1) Find an appropriate air quality objective. The WHO Expert Committee's goals are generally accepted in the UK and in future the EEC's standards or the Royal Commission on Environmental Pollution's guidelines may be used. You can refer to standards or objectives used in other countries—though some people in the UK will question whether other countries' decisions are appropriate here. This will be especially true of the Soviet standards which aim to give much fuller protection than pollution authorities in western Europe and the US normally claim is 'practicable'.

Most countries and international bodies only set objectives for a handful of the most common air pollutants (the Soviet Union is an exception); if there is no objective for the pollutant in which you are interested, go directly to the sources listed in chapter 2.

(2) Make a point of looking up the aims of any air quality objective you use. At first sight they may seem to be setting completely 'safe' levels, but as is explained in chapter 1, they often aim to give the maximum degree of safety that can be achieved *at a reasonable cost*.

If an objective claims to avoid 'all ill-effects' try and find out how much is included in this phrase. You may have to spend a little time looking at the evidence on which the objective is based; usually this will have been collected together as an 'air quality criteria' report. You can normally assume that the objective will have been set at concentrations below those that have so far been found to cause or contribute to death or severe illness. It may not necessarily cover concentrations that contribute to minor ailments or make existing conditions slightly worse. Do not assume—unless a specific claim to the contrary has been made—that an objective designed to protect human health will also prevent nuisance to the community or damage to the natural environment.

(3) Pay special attention to the 'averaging time' on which an

objective is based. Pollution monitoring will usually show the average concentration of a substance in the air over a given period—anything from an hour to a year. The longer the averaging time, the lower will be the concentration detected, because short-term peaks are averaged out (see pages 105–6).

When you compare air pollution concentrations to an air quality objective, be sure to check that the averaging time of the monitoring results is the same as the averaging time of the objective, in other words, compare daily smoke levels with the WHO goal for average *daily* concentrations but not with the goal for the *annual* average.

When a standard is in more than one part, for example if different concentrations are set for different averaging times, air quality is satisfactory only if it meets *all parts* of the standard—but not if it is within, say, the goal for the annual average concentration but exceeds the goal for the daily concentration.

5 Air Pollution Monitoring

Each of the subjects taking part in the experiment was given a pair of socks to be worn on two days. . . . All the socks were washed and the amount of lead normally present was measured. . . . Analysis of the socks found an average lead level (measured in mg/sock) of 7·3 for administrative staff, 17·4 for supervisory staff, 53·9 for manual staff inside the factory, 39·8 for manual staff outside the factory, 0·16 for the control group. . . . It would seem that wherever work is done in the factory, lead is picked up on clothes even by those whose work does not take them into the areas where lead would be expected to be present.

Municipal Engineering (28 March 1975)

Summary

Air pollution monitoring surveys are usually designed to show one of three things: (1) which pollutants are in the air and whether their concentrations are high enough to cause damage to human health or the environment, (2) how concentrations of pollution have changed with time and how successful pollution control policies have been, and (3) how much ground level pollution is caused by particular sources.

Information from a survey designed to answer one of these questions will often be of little use in dealing with any of the others. In particular, national surveys of pollution trends are generally unable to demonstrate how much pollution a particular factory near one of the monitoring sites has added to the air.

When interpreting monitoring results, check that (a) the sites have been chosen to provide the kind of information you need, (b) the period over which results are averaged is not so great as to smooth out important but short-lasting peaks of pollution, (c) enough samples to give a fair picture have been taken, (d) any unusual weather conditions or changes in emissions have been taken into account, and (e) you are aware of possible inaccuracies or errors.

Monitoring programmes

Two pollutants, smoke and sulphur dioxide (SO_2), have been systematically monitored throughout the UK for many years. They are amongst the most common air pollutants, since they are produced whenever fuel is burned; their presence is taken as a sign that other pollutants produced by combustion are also in the air.

Local authorities carry out most of the monitoring, though in some areas it is done by companies or other organisations. The results, covering about 1150 town sites and 150 country sites, are compiled by the Department of Industry's Warren Spring Laboratory and published as the *National Survey of Air Pollution*.

The Laboratory's regular publications show how successfully smoke and sulphur dioxide pollution has been reduced by the Clean Air Acts and by the change from coal to more convenient smokeless fuels. By 1973-74 the average concentration of smoke in the UK was one-third of its level in the early 1960s and SO_2 concentrations had fallen by 40 per cent.[1]

Many local authorities also measure the amount of *grit and dust* deposited from the air. Most of this solid material is made up of ash, unburnt solid fuel, dust blow from roads or the land and fallout from various industrial processes.

The heavier particles fall out of the air and are often the cause of public complaint. Fallout is collected in some 600 'deposit gauges' throughout the country and the results have shown that there has been no significant decrease in the amounts of solids deposited since the 1960s. While smoke and sulphur dioxide concentrations tend to be higher in winter than in summer (because more fuel is used for heating and because meteorological conditions are often unfavourable for pollution dispersal in winter), grit and dust concentrations show little change between seasons, as relatively little of it comes from domestic heating.

Results from the deposit gauges only tell part of the story. The gauges collect the dust that is likely to cause nuisance, but they do not measure the concentrations of the very tiny particles that remain suspended in the air. These fine solids are potentially the most dangerous since their small size allows them to penetrate into the lungs.

A number of local authorities and research bodies monitor other air pollutants; *lead* levels in the air are measured by nearly 50 authorities, and other *toxic metals* are also often monitored. This is sometimes done by measuring the rate at which specially prepared bags of moss absorb metals from the air. The method not only gives some idea of how much of these metals is in the air but also indicates how quickly they are being taken up by vegetation and how much may be reaching man through foodstuffs such as vegetables, meat and milk.[2]

In 1974, an official review of monitoring in the UK recognised that the scope of air pollution monitoring in the past had been limited and announced several new surveys.[3] A network of 20 new *metal* monitoring sites would be set up over the country, and 20 other sites would measure the concentrations of *acidic particles and aerosols* (mists of tiny droplets in the air), such as the highly toxic sulphuric acid that may be formed in the air from sulphur dioxide. Three of these sites would also measure concentrations of *nitrogen oxides, ozone* and *hydrocarbons*—substances involved in the production of highly irritating and damaging 'photochemical smog' from motor vehicle exhausts. (A separate survey of traffic pollution in five towns has been in operation since 1972.)[4] A preliminary report on the siting of the new monitoring sites has also been published recently.[5]

Where to find monitoring results

NATIONAL SURVEYS

Smoke and sulphur dioxide concentrations, measured in the *National Survey of Air Pollution*, are published by Warren Spring Laboratory (WSL)—see[6] for address.

To use the results from this survey you need to look at two separate publications (1) the *Directory of Sites*,[6] which gives the address of each site (see Figure 5.1) and a code which roughly classifies the kind of area surrounding it, and (2) annual tables of results,[7] which give the name of each site (but not its location), the average ('mean') concentration of smoke and SO_2 at the site for each month, season and the whole year, and the number of days in each month when pollution concentrations exceeded certain levels (see figure 5.2). This information may also be published by some of the local authorities who operate *National Survey* sites.

The concentrations of pollution found on any individual day are not normally published, though you should be able to find this information by contacting the WSL or the organisation that operates the monitoring site.

A discussion of monitoring results collected in the *National Survey* between 1961 and 1971 has been published by the WSL in a series of five reports.[8] Although the results are now out-of-date, the reports contain very useful discussions on local factors affecting pollution in each town. They also include (a) maps showing the location of all monitoring sites, industrial sources of pollution and smoke control areas, and (b) tables and graphs showing trends in pollution over the decade studied (see figures 5.3 and 5.4).

Site Name and number		Co-operating Body	Site Code	Observations Started	Observations Stopped	Address of Site	Grid Reference
Bedford	2	B	D1	1948	1961	P.H. Dept., Town Hall	TL/049 496
	3		D1	1960	1970	Public Library, Harpur Street	TL/049 498
	4		B3/E	1960	1965	91 Queens Drive	TL/070 513
	4		B3/E	1968		91 Queens Drive	TL/070 513
	5		C2	1961		Wm. Allen's, Queens Park, 3 Ford End Road	TL/041 496
	6		A2	1961		C.D. Centre, 22 Albert Street	TL/047 504
	7		B2	1961		Welfare Clinic, 29 Barford Avenue	TL/058 487
	8		B3/E	1961	1963	6 Merlin Gardens, Brickhill	TL/053 516
	9		B3/E	1963	1968	Clinic, Linnet Way, Brickhill	TL/052 516

FIGURE 5.1

Bedford monitoring sites, extract from *Directory of Sites* used in the investigation of air pollution

Table 1 Smoke and Sulphur Dioxide   April 1972–March 1973
Units – micrograms/cu m

Site Name and Number	Class Code			Smoke AV	Smoke HD	No of Days exceeding 250	No of Days exceeding 500	No of Days exceeding 1000	SO_2 AV	SO_2 HD	No of Days exceeding 250	No of Days exceeding 500	No of Days exceeding 1000	Smoke/SO_2 Ratio
Bedford 4	B3/E	A	72	N	185				N	68				N
		M	72	9	21				26	68				·35
		J	72	8	19				24	62				·33
		J	72	7	17				15	44				·47
		A	72	8	24				20	54				·40
		S	72	68	183				35	62				1·94
		O	72	N	148				N	96				N
		N	72	71	123				94	164				·76
		D	72	N	107				75	134				N
		J	73	68	119				84	150				·81
		F	73	41	120				91	166				·45
		M	73	34	101				80	160				·43
		Mean S		21					25					·84
		Mean W		52					82					·63
		Year		37					53					·70

FIGURE 5.2

Smoke and SO_2 observations in Bedford, 1972–73; extract from *The Investigation of Air Pollution—Observations of Smoke and SO_2*

Site	Short description
3	Public Library; Harpur Street, in the administrative centre, behind the shops.
4	Queens Drive Clinic; low-density housing at edge of town, overlooking the old town. Smoke control.
5	W. H. Allen (Pumps) Ltd; heavy industry, dense housing.
6	Albert Street; congested high-density terraces, being opened up by demolition.
7	Barford Avenue Clinic; inter-war dense council estate on the valley floor.
8	Merlin Gardens; as No. 4 but higher.
9	Linnet Way Clinic; as No 4 but higher. Smoke control.

Average pollution, $\mu g/m^3$, winter ended March:

	1962	1963	1964	1965	1966	1967	1968	1969	1970	1971
Smoke										
3	147	168	108	94	–	74	95	85	109	–
4	120	117	88	71	–	–	–	–	46	–
5	173	182	126	–	91	–	–	77	–	69
6	–	233	–	145	–	98	108	–	–	76
7	–	193	132	102	–	94	109	94	85	73
8	–	88	–	–	–	–	–	–	–	–
9	–	–	69	55	–	44	54	–	–	–
Sulphur dioxide										
3	240	264	218	213	–	183	181	240	163	–
4	107	134	99	55	–	–	–	–	109	–
5	191	217	181	–	149	–	–	173	–	91
6	–	183	–	125	–	83	67	–	–	96
7	–	198	162	159	–	113	164	180	128	103
8	–	112	–	–	–	–	–	–	–	–
9	–	–	34	81	–	82	98	–	–	–

FIGURE 5.3

Smoke and SO_2 trends in Bedford, 1961–71; extract from *National Survey of Air Pollution 1961–71, Volume 1*

The WSL has also used the *National Survey* to produce contour maps of the UK showing areas in the country where smoke and SO_2 exceeded certain levels.[9]

Results of deposit gauge measurements of grit and dust fallout are published separately by Warren Spring Laboratory.[10]

LOCAL SURVEYS

Your local authority will probably be able to tell you whether it, or any other organisation, is measuring air pollution concentrations in your area, and where the results of this monitoring can be found. The WSL may also be able to tell you about local surveys.

Details of universities and other bodies carrying out research into air pollution (which may involve special monitoring surveys) are given in the Department of the Environment's *Register of Research*[11] and in the yearbook of the National Society for Clean Air.[12]

Meteorological data about wind speeds and directions and temperature inversions may be needed if you are trying to use monitoring data to pinpoint the source of a pollutant. Details are published by the

Meteorological Office (London Road, Bracknell, Berkshire, telephone: 0344 20242).

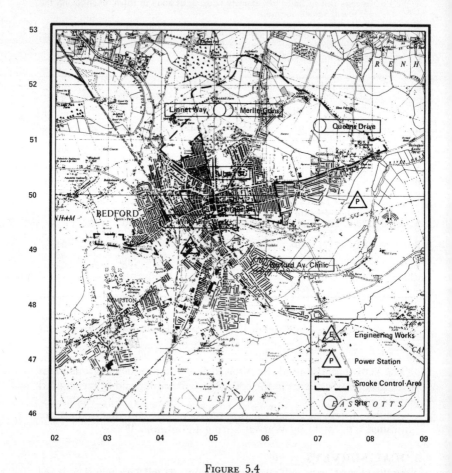

FIGURE 5.4
Monitoring sites, factories and smoke control areas in Bedford; extract from *National Survey of Air Pollution, 1961–71*

Pinpointing the source of pollution

Pollution monitoring may be done either to give a picture of pollution levels in a particular area or to measure the impact of an individual factory. **Surveys that have been designed to give general information about pollution levels cannot normally also be used**

to pinpoint the source of a pollutant or to trace emissions from a particular factory.

To do this, sites would have to be located all around the factory being investigated, measurements would have to be made continuously—or at least very frequently during the day—and the results studied with records showing the direction of the wind at the time of each measurement.

The *National Survey* of smoke and sulphur dioxide, and the measurement of grit and dust fallout by deposit gauges cannot normally be used in this way. The deposit gauges show the total amount of solids deposited in a month. Now if the wind had been blowing constantly in one direction for one month and then constantly in another direction for a different month, it might be possible to tie high readings in with periods when the wind blew directly from the direction of a dust-producing factory. In practice, the wind changes direction all the time, and results from days when the wind was coming from the source of the dust are mixed in with the rest of the month's dust, making it difficult or impossible to demonstrate any connection. (The Central Electricity Research Laboratory has now developed a device which will provide more information about the sources of dust pollution. It measures the amount of dust in the air—not that which falls out—and separates out the dust originating from different directions.[13])

Smoke and sulphur dioxide monitoring from the *National Survey* is a little more useful, since readings are taken every day to show the average concentration over a 24-hour period. Even so, the wind may be changing direction constantly during the day and there are often changes in pollution concentrations at night time (when factory emissions may cease and when temperature inversions tend to prevent the dispersion of emissions) which are concealed by a single 24-hour average figure.

On top of this, *National Survey* monitoring sites have usually been chosen to give a picture of smoke and SO_2 levels in representative districts in each town (such as residential area, smoke control area, town centre, industrial district[14]) and not to show the effects of a particular factory. *Do not be misled by the fact that sites are often located on the premises of power stations, gas works or other factories.* Monitoring equipment has usually been put in these places to show typical pollution levels in an industrial area. Although they will pick up some low-level emissions from the factory itself they will not be exposed to the bulk of the factory's pollution. Under normal weather conditions most of the factory's chimney emissions will have their maximum impact at a distance of about 15 times the height of the chimney; very little will come down on the factory premises (see page 78).

Separating industrial and domestic pollution

Although results from the *National Survey of Air Pollution* cannot normally be used to show how much a particular factory has added to smoke and SO_2 pollution levels, they can, in very broad terms, reveal how much of these pollutants originates with industry and how much with domestic fires. This is done by looking at two ratios: the ratio of winter pollution to summer pollution, and the ratio of smoke to SO_2 pollution.

THE WINTER/SUMMER RATIO

If the only source of smoke and SO_2 pollution in an area is from a source that operates throughout the year, for example, industry or traffic, pollution levels will remain more or less constant over the year. There may, however, be a slight increase in the winter months because inversions, which interfere with the normal dispersion of pollutants, are more common then. On the other hand, if pollution is caused by fuels consumed in heating appliances the concentrations of smoke and SO_2 will rise markedly during the cold months.

The annual tables of results published by Warren Spring Laboratory show the average concentrations of smoke and SO_2 over the summer and winter seasons. Results from the 'Bedford 4' site in figure 5.2 show that during the winter the concentrations of both smoke and SO_2 were more than double the summer concentrations, a clear sign that domestic heating (or heating of factories and offices) contributed large amounts of pollution during the winter months.

A large increase in concentrations during the winter does not necessarily mean that heating is the only—or even the main—source of pollution.

Ask yourself what is causing the summer pollution. A large part of it may come from industry—although even more may be added to it from domestic heating during the winter. The industrial component may have a greater impact in the winter than in the summer because it may be less easily dispersed.

At sites where there is little change in readings between the seasons you can assume that pollution from non-heating sources is responsible. This does not automatically identify industry as the polluter. Any other non-seasonal source of smoke and SO_2—for example traffic or trains—could be the cause.

THE SMOKE/SO_2 RATIO

The amounts of smoke and sulphur dioxide given off by different fuels vary depending on the sulphur content of the fuel and the efficiency of the furnace or fire in which it is burned. (Information about the efficiency of different appliances is shown in table 3.2 on page 75.)

(a) *Domestic coal* produces slightly more smoke than SO_2, about 0·5 kg (1·25 lbs) of smoke for every kg of SO_2. All other industrial and domestic fuels produce more SO_2 than smoke.

(b) *Solid smokeless fuels* are coke-like substances, specially produced from coal to burn smokelessly. They contain slightly less sulphur than normal coal but still emit relatively much more SO_2 than smoke—about five times more.

(c) *Industrial coal* gives off very much less smoke than coal burned in domestic fires—not because the coal is very different, but because industrial boilers burn coal far more efficiently than domestic fires. Coal burned in industrial boilers produces over ten times more SO_2 than smoke, depending on the efficiency of the boiler.

(d) *Oil* is normally burned very efficiently and produces little smoke, but it contains a substantially higher proportion of sulphur than coal or solid smokeless fuel. Oil releases about 50 times more SO_2 than smoke.

(e) *Natural gas* burns smokelessly and contains no sulphur at all.

If we assume that the concentrations of smoke and SO_2 reaching the ground are in the same proportions as they are in emissions, then the ratio of smoke to SO_2 found in the air will give some idea of the kind of source that produced the emission. For example, if the smoke/SO_2 ratio is between about one and two, then emissions have probably come from domestic fires burning coal. If the ratio is significantly less than one (in other words, the concentration of SO_2 found in the air is greater than the concentration of smoke), then the pollution has been caused either by the burning of domestic solid smokeless fuel or by coal or oil used in industry.

Very often pollution at ground level comes from a variety of sources—both domestic and industrial. So do not expect to find that either industry or home fires are solely responsible. Table 5.1 will give you some idea of the smoke/SO_2 ratios you would find if pollution did come exclusively from one source.

TABLE 5.1
The ratio of smoke to SO_2 produced by burning various fuels

Fuel	Smoke/SO_2 ratio
Domestic coal	1.25
Solid smokeless fuel	0·2
Industrial coal	0·1
Oil	0·02

One factor that can sometimes complicate the use of this ratio is smoke from traffic, which is much denser than smoke from other fuels. Traffic smoke stains the filter paper used to measure smoke density much more than the equivalent weight of smoke from industry or domestic fires, so traffic pollution will tend to increase the smoke/SO_2 ratio.

The monthly and seasonal smoke/SO_2 ratios for each site in the *National Survey* are published by Warren Spring Laboratory in the tables of monitoring results (see figure 5.2). The Laboratory has also published graphs showing how the ratio can be used to tell you *very roughly* how much pollution has come from domestic coal and how much from all other sources.[15]

If you are looking at changes in either the smoke/SO_2 ratio or the winter/summer ratio, remember that these only show the relative amounts of the two factors—not the absolute levels. Any change in the ratio could be caused by a fall in one factor or by an increase in the other.

Interpreting results

The rest of this chapter explains some of the key questions you should ask when looking at monitoring results. The points are illustrated using sampling results from the National Survey, but the principles apply to data from any air pollution survey.

Samples

Do you have enough samples to give a fair picture?

Ground level concentrations of pollution can vary by large amounts from hour to hour and from day to day depending on the wind and other weather conditions. Table 5.2 shows how concentrations of smoke and SO_2 near Doncaster differed on two days in the same week in 1971.

TABLE 5.2

Changes in pollution concentrations near Doncaster on two days during the same week in 1971[16]

	11 November		17 November	
	smoke	SO_2	smoke	SO_2
	($\mu g/m^3$)		($\mu g/m^3$)	
Site 1 (Barnby Dun)	471	360	35	92
Site 2 (Sprotbrough Clinic)	558	420	64	58
Site 3 (Alston Road)	246	159	22	69

Air Pollution Monitoring

One or two days' worth of sampling results may give you a very distorted picture of pollution concentrations during the week or month, especially if weather conditions have altered during this period. In the same way, annual pollution cannot be assessed by looking at one or two months' results. The more variation there is in weather conditions over the period you are looking at, the more samples you will need to get a fair picture.

Warren Spring Laboratory does not publish an average figure for pollution concentrations during the winter or summer seasons unless it has reliable figures for about 60 per cent of days, with no gap of more than 14 days' results missing. It does not publish an average for the whole year unless it has at least 220 days' readings. This is probably a good guide to the minimum number of readings you should aim to use to give a reliable average figure for any period.

Averaging time

Are the monitoring results you are using averaged over such a long time that they conceal the most serious pollution episodes?

If the data you are using gives a single figure for the average pollution over 24 hours (as the *National Survey* does) you will not know anything about short-term high peaks of pollution which may have lasted only an hour or so. The 'peaks' will be balanced out by the 'troughs'—the times when pollution concentrations were very low—and the average will be somewhere in between. These short-term high bursts of pollution may be very harmful to health—but be completely unnoticed if only daily readings are taken. Hourly measurements of pollution concentrations in Sheffield have shown that on one badly polluted day:

> for 7 hours SO_2 levels exceeded 1000 $\mu g/m^3$ (from the standpoint of health a serious level) and for 4 consecutive hours exceeded 1800 $\mu g/m^3$, reaching 3200 $\mu g/m^3$ for one hour. Yet on a daily sampler at the same site the mean daily assessment was no more than 777 $\mu g/m^3$. Daily means must therefore be accepted with some reservation and a sense of inadequacy for they only indicate part of the evidence.[17]

If your figures show average pollution concentrations over the month, or even the year, you will miss out even more of the very high concentrations, including those that persisted over several days or even weeks.

Table 5.3 shows the results of carbon monoxide monitoring in Chicago in 1962. The concentration was monitored continuously over the year and the results were calculated using different averaging times. As the averaging time was increased, the maximum concentration observed dropped.

TABLE 5.3

Maximum concentrations of carbon monoxide observed when different averaging times are used on the same set of results[18]

Averaging time	Maximum concentration (parts per million (ppm))
5 minutes	50
1 hour	36
8 hours	22
1 day	19
1 week	13
1 month	10
1 season	9
1 year	8

Automatic monitoring equipment that will show the concentration of pollutant at any instant is available—though it is much more expensive than the conventional equipment. If you feel that daily average results are concealing harmful short doses of pollution, ask your council to monitor using continuous monitoring equipment.

Pollution concentrations

Are you looking at the worst pollution concentrations as well as the average?

Pollution in the winter is often greater than in the summer because temperature inversions, which prevent the normal dispersion of pollutants, are more common. At the same time pollution from domestic heating is also greatest in the winter. So look at winter pollution figures rather than averages for the year as a whole. Also look at the highest monthly and daily concentrations and not just at the averages. These high concentrations may occur only for a tiny fraction of the time, but they may be the most dangerous.

The tables of results published by Warren Spring Laboratory for smoke and SO_2 concentrations will help you locate these high peaks. Winter average concentrations are shown ('Mean W' in figure 5.2) and so are the highest daily concentrations found in each month ('HD' in figure 5.2). The tables also show the number of days when concentrations exceeded certain levels. Ask for the same kind of details when you are looking at results from the monitoring of other pollutants.

Monitoring sites

Have monitoring sites been put in the right places?
Ideally, pollution should be monitored at the places where (a) the greatest concentrations of pollution are likely to be found, and (b) the greatest numbers of people or the most vulnerable sections of the population or environment are found. In practice, monitoring sites are often badly located. According to Warren Spring Laboratory:

> Instruments can only be operated where the owner of the premises is willing to have them, where they are safe from interference, and where the local public health officers have free access to them. This practically limits the choice of sites to official buildings of one sort or another and the availability of such buildings affects the pattern of measurements in most towns.[19]

Look up the position of monitoring sites on a map of the area. (Maps showing the position of sites used in the *National Survey of Air Pollution* have been published.[8]) Equipment to measure pollution from domestic fires will need to be quite close to residential areas because house chimneys are relatively low and emissions are not widely dispersed. On the other hand, factory pollution will be spread further afield because discharges are made through tall chimneys. Some more localised factory pollution may be caused by emissions from stockpiles of dusty materials or by gases extracted from the factory atmosphere and discharged at roof level.

Try and examine the position of a monitoring site to see how well it represents the surrounding area. A site located in a smoke control area will not give readings of pollution in surrounding areas where householders may be allowed to burn coal. The site may be influenced by untypical factors which bias the results; for example, if it is close to a boiler chimney, or in an enclosed yard, or on a busy road. (Daily smoke concentrations on a busy street in Hackney have been found to be 200 µg/m³ higher than at other sites away from traffic.[20] This may tell you what concentrations pedestrians on this particular street are exposed to, though it may not represent pollution in the surrounding area.)

Warren Spring Laboratory publish advice on how to choose monitoring sites that represent the surrounding area.[21] If the site has not been carefully chosen you may not be able to draw conclusions about pollution levels in the areas near it. However, it may still accurately reflect *changes* in pollution concentrations even though it does not give any absolute readings of the actual concentrations.

108   *The Social Audit Pollution Handbook*

Weather conditions and emissions

Were weather conditions and emissions typical at the time of monitoring?

It is important to know whether monitoring results were obtained during typical conditions or whether they have been influenced by unusual factors that only occur rarely. Pollution concentrations may be higher than normal during periods of very low wind speed or temperature inversion. They may be lower than normal during a warm winter (when little fuel is used for heating) or when emissions are temporarily reduced (for example, during a strike or the factory's annual holiday).

Short-term factors of this sort may account for changes which might otherwise seem to be caused by better—or worse—control over factory emissions. Looking at several years' worth of monitoring results—if they are available—will usually help you tell the difference between real trends in pollution and short-term fluctuations.

Errors and Inaccuracy

If you are comparing pollution concentrations at different sites, have you taken all possible errors in measurement into account?

Monitoring techniques do not give absolutely accurate readings: the methods themselves always involve some inaccuracy. In the case of the SO_2 measurements taken for the *National Survey*, the method itself is only claimed to be accurate to within ten per cent of the actual concentration.[22] Thus, if the actual concentration of SO_2 in the air is 100 μg/m³ (micrograms weight of SO_2 per cubic metre of air) the readings could show anything from 90 μg/m³ (ten per cent lower) to 110 μg/m³ (ten per cent higher). It would be quite possible for two different instruments to measure exactly the same sample of air but give results differing by this amount.

If the difference in readings from different sites is small, find out what the accuracy of the method of measurement is. The difference could be explained by the inaccuracy of the technique.

Example:

Two instruments measure pollution levels to within an accuracy of ten per cent. The first shows a reading of 220 μg/m³ and the second of 250 μg/m³. Is this a significant difference?

Add ten per cent of the lower reading to that reading (220 + 22 = 242). Subtract ten per cent of the higher reading from the higher reading (250 − 25 = 225)

Thus, taking the possible inaccuracy of the method into account it is possible that the concentration of pollutant at the first site is actually *greater* than the concentration at the second site—and not *lower* as originally appeared.

On top of the inaccuracy inherent in the method, results may be biased by a whole range of errors caused by faulty equipment or handling. Insects have been found living in the inlet tubes to monitoring equipment where they tend to block the tube off, reducing readings. In other cases the equipment may not be put together correctly. One of the organisations responsible for operating *National Survey* sites has reported that:

> after repeated instruction some junior staff and students changing the apparatus seem to be more interested in breaking records for the time taken in changing the sets rather than in seeing that the filter papers are 'smooth side down' and placed centrally in the clamps.[23]

Errors of this sort, and the difficulties in finding sites that are representative of the general area, have led one observer, Professor Scorer of Imperial College London to issue this warning about results from the *National Survey:*

> What we have to fight is the *assumption* that the observations are correct. I'm not disputing that most of them could be fairly valid, but if any of them are incorrect we almost certainly don't know which they are. Nor have we any idea what proportion is reliable, and we don't use the results to the accuracy with which the measurements are made, or are claimed to possess.[24]

Note that there is a difference between the *inaccuracy* of a technique and *errors* that may be made by mishandling equipment. The term 'accuracy' refers to the precision of the method itself: however carefully the equipment is used the accuracy of the method cannot be improved. However, 'errors' can be eliminated altogether. The more readings that are put together to give an average, the more likely it will be that the errors will cancel each other out.

Unless you know the accuracy of a method you cannot make very precise judgements about whether one site is slightly more or less polluted than another—though where the difference between sites is relatively large you do not need to worry about the method's accuracy.

Using sampling data

Are you trying to prove that a particular factory does or does not add significantly to the levels of pollution in the area by using unsuitable sampling data?

If you are, you are in distinguished company—but it is still wrong to do it. For example, at a planning enquiry held in 1970 to decide whether a Coalite smokeless fuel factory should be built at Rossington near Doncaster, a consultant engaged by Coalite and Chemical Products Ltd. analysed the *National Survey* results for three areas where Coalite factories already operated and concluded:

> In my opinion the Coalite Works at Bolsover, Askern and Grimethorpe as operated during the years 1964 to 1968 appear to have had no appreciable effect on the pollution of the air near ground level in the surrounding areas by either smoke or sulphur dioxide.[25]

The following argument (which was applied in turn to each of the three existing Coalite works) was used to support this conclusion:

(1) The consultant examined monitoring results from six *National Survey* sites in the Bolsover area for evidence of pollution caused by the existing Coalite plant there. But no evidence was given to show that any of these sites had been placed in a position where it would record the maximum impact of the factory's emissions.

(2) By comparing the concentrations of smoke and SO_2 recorded at these sites during the winter months with the summer levels, the consultant demonstrated that in winter considerable amounts of pollution were caused by domestic fires. However, this did not demonstrate that industrial pollution was insignificant.

(3) Finally the consultant compared the concentration of SO_2 at the six Bolsover sites *averaged out over five years* with the concentrations in the whole country. The range of results found at Bolsover was lower than the range for the rest of the UK, and he concluded: 'it is thus clear that there is no evidence that the discharges from the Coalite works . . . have had any appreciable effect on the pollution of the air at the measuring sites'.

However, Bolsover's pollution could have come entirely from the Coalite works and still be lower than the national average. It would be equally possible for Bolsover to have double the national average—with all the pollution coming from sources other than the Coalite works.

Even if such a comparison had been valid, all but the most enormous differences between sites would have been averaged out and lost by looking at the five-yearly average results. (In response to these comments, the consultant subsequently noted: 'Obviously I could only use the known results of measurements of pollution of the air in areas surrounding existing Coalite Works. To obtain detailed evidence of the kind you consider necessary for decision at such an inquiry, would obviously mean careful verification of selected measuring sites, with more or less continuous records of results throughout each day over at least three years.')

In short, it may be very misleading to try and quantify the exact impact of a factory's emissions by relying on a monitoring programme that has not been designed specifically for the purpose.

Using monitoring results

(1) Try to get the help of someone with experience in interpreting monitoring results. This chapter explains the main principles to follow in handling monitoring data, and it shows you how to think about the information. But reading it will not make you an expert—though it will help you question the way other people use monitoring results to prove particular points. You will find it helpful to talk over a local problem with someone who knows why monitoring sites have been put in particular positions and is familiar with the strengths and weaknesses of the methods used. The best place to get advice is from the organisation carrying out the monitoring. Often this will be the local authority, though you may also be able to consult Warren Spring Laboratory, the Alkali Inspectorate, the company producing the pollution, or research staff at universities or scientific institutions.

(2) Decide what you want to learn from the results of monitoring. This might be (a) how much a particular factory is adding to air pollution levels, (b) whether existing concentrations of pollution are likely to cause damage to human health or to the environment, and (c) whether pollution control policies have succeeded in reducing air pollution levels.

(3) Check whether existing monitoring is capable of giving you the information you need. Find out whether the right pollutants are being monitored and whether measurements are being made at suitable positions. The choice of monitoring sites will determine whether the measurements can be used to check on the impact of individual sources or whether they only give more general information about pollution in the area. Find out why particular sites have been chosen for measurement. Ask whether these sites represent points where (a) the maximum concentrations of pollution in the area are likely to be found, (b) the most sensitive sectors of the population or environment are situated (a hospital or a newly planted forest, for example), or (c) large numbers of people live or work, (d) or it happens to be convenient to leave monitoring equipment, although the sites themselves are not particularly representative of the surrounding areas.

See that you have a reasonable number of samples from each site and that the results are not averaged out over such a long period that you overlook important short-term variations. If readings show the average monthly (or daily) concentrations and you need the average daily (or hourly) concentrations, ask your local authority to consider buying continuous, or at least better, monitoring equipment.

If you are going to compare readings taken at different sites, you will need to know the accuracy to which the instruments are capable of measuring pollution. Do not try to make any very precise comparisons without taking the possible inaccuracy of the measurements into account.

(4) If you are trying to find out how much pollution is coming from a particular source check whether there are any other possible sources of the same pollutants in the region. If there is only one factory that produces a certain pollutant in the area, the source will be obvious. Sometimes you will find that several different factories could be causing the pollution, and to pinpoint the culprit you will need to look at wind directions at the times measurements are made. Alternatively, it may be possible to collect a sample of a solid pollutant and have it analysed by the local authority's analyst or by a public analyst (who will charge you a fee). The shape or composition of the particles may identify the source.

If you can identify the amount of pollution at ground level caused by one factory you may want to consider whether (a) this is likely to be causing a hazard to health or damage to the environment, (b) high measurements found when the wind is in a particular direction can be linked to any complaints of nuisance that may have been made, and (c) the impact of the factory's emissions conforms to any predictions or promises made before the factory opened, for example at a planning inquiry.

Find out from the factory or from the pollution control authority whether high environmental concentrations have been caused by normal emissions or whether exceptional circumstances (such as a breakdown of control equipment) were to blame. Then enquire what steps are being taken—or could be taken—to reduce emissions, and what the costs and likely benefits will be.

(5) If you are trying to explain any change in pollution concentrations over time see whether it can be linked to changes either in emissions or in meteorological conditions which might affect the dispersal of pollutants. Changes in emissions from industry might be caused by the opening or closing of factories, a change in method or volume of production, the use of a new fuel, shutdowns caused by industrial disputes or holidays or by improved (or deteriorating) pollution control equipment. Domestic emissions may be influenced by a change in the fuel used to heat houses, especially where a new smoke control area has been created. The volume of domestic emissions may change from year to year—or from day to day—with the weather. A stretch of warm weather when it had previously been cold may be the reason for a drop in concentrations of pollution. Conditions of dispersal will change not only from day to day but also from year to year; the most important factors to look at are wind speed and direction and the occurrence of temperature inversions.

(6) If you want to know whether pollution levels may be hazardous find out what concentrations and periods of exposure have been found to cause specific effects (see chapter 2) and compare them with the concentrations measured in the area at which you are looking. The averaging time used in the studies of hazardous effects must be the same as that used in your local monitoring: compare hourly, daily, annual etc. average concentrations in both cases. Pay special attention to the effects of short exposure to the highest concentrations revealed by the monitoring programme.

Inadequate monitoring may conceal hazards either because the averaging time is too long to pick out short-term high readings or because readings are not taken at the right places. Concentrations in unmonitored areas may be greater than in those places where monitoring equipment has been put; so do not assume that because measurements have not been taken that there is no hazard.

Water Pollution

The water pollution section of this Handbook is divided into three chapters.

River Pollution Law and Standards (chapter 6) describes the legal controls over discharges of effluent to rivers or public sewers. It explains how you can find out (a) the standards applied to discharges from individual factories or sewage works, (b) the quality of discharges from these sources, and (c) the likely impact of discharges on concentrations of pollution in the river.

Water Quality Objectives (chapter 7) describes the systems used to classify rivers according to their chemical quality and their ability to support river life. It explains how river quality objectives are set and which standards are used to ensure that a river can support fish and that drinking water is fit for human consumption.

River Pollution Monitoring (chapter 8) tells you how to find out how much pollution is present in a river. It explains how to use river pollution monitoring results to discover (a) where toxic substances in the river are coming from and how much individual factories or sewage works contribute to river pollution, (b) whether changes in river pollution are due to pollution control policies, and (c) whether a river is likely to be poisonous to fish or hazardous to man. Sources of information about the toxicity of water pollutants are given in Part 9 of **Investigating Hazards** (chapter 2).

6 River Pollution Law and Standards

Summary

Industrial effluent may be discharged directly into rivers or first into sewage works for treatment with domestic sewage. No effluent can be discharged to a river—either from a factory or from a sewage works—without the consent of the water authority which controls the quality of discharges by attaching conditions to its consent.

When the 1974 Control of Pollution Act is brought fully into force, existing consent conditions will be reviewed and conditions may be made more stringent or be relaxed, depending on the use to which the river is put. Many of these consent conditions will be toned down, so that discharges which previously exceeded 'unrealistic' consent conditions, and which could be improved only by considerable spending, will become legal.

Under the Control of Pollution Act, members of the public will be given access to full details about the quality of effluent discharges from individual factories and sewage works. They will also have the right to prosecute dischargers who exceed their legal consent conditions.

Water authorities

Industry's liquid wastes are known as 'trade effluents'. They can be discharged in two ways:

(a) directly into a 'watercourse' such as a river, estuary or the sea;

(b) into a public sewer for treatment at a sewage works and then discharged with the sewage effluent into a watercourse.

This chapter deals only with trade and sewage effluents discharged into freshwater rivers. If you are interested in discharges made to tidal rivers, estuaries or the sea you will find an account of the relevant laws in the books listed in reference.[1]

No trade effluent can be discharged into a watercourse without the consent of the regional water authority (or, in Scotland, either the river purification board or the local authority). Since 1974, the water authorities have been responsible for controlling river pollution and sewage disposal—functions previously shared between the old river authorities and local authorities. Addresses of water authorities are shown on page 173. Details of their local offices can be found in *Who's Who in the Water Industry*.[2]

Effluent discharged into a river

At the time of writing (March 1978) trade effluent discharges to rivers were controlled under the Rivers (Prevention of Pollution) Acts 1951 and 1961. However, these Acts were due to be replaced by the 1974 Control of Pollution Act which was shortly to be brought fully into force. This chapter describes the system laid down in the Control of Pollution Act. You can find out from your water authority whether the water pollution sections of the 1974 Act are actually in force.

THE NEW SYSTEM
The Control of Pollution Act forbids the discharge of any poisonous or polluting material (or anything that would aggravate pollution from another source) into a river without the consent of the water authority.[3] 'Causing or knowingly permitting' an unauthorised discharge is punishable by up to two years' imprisonment and/or a fine.

The water authority regulates discharges by attaching conditions to its consent. These may limit the nature, composition, temperature, volume and rate of the discharge and require the discharger to sample the effluent regularly and provide the water authority with full results.[4]

Water authorities will be required to review their consent conditions from time to time, though they will not normally be able to alter a consent before a 'reasonable period' of time (which must be at least two years) has passed.[5] However, a water authority will be obliged to change or even cancel a consent—if necessary before the 'reasonable period' has ended—if it believes that a permitted discharge: (a) may harm human health;[6] or (b) has injured plant or animal life in the river.[7] (If this happens, the water authority will be required to take steps to repair the damage and if possible to restore the river to its former state.)

A water authority that changes a consent prematurely may have to pay compensation to the discharger—unless the change is necessary because of altered circumstances.[8]

THE OLD SYSTEM

Until the Control of Pollution Act comes fully into force, water authorities will continue to control effluent discharges through similar, but less extensive, powers to issue consents under the Rivers (Prevention of Pollution) Acts.

Existing consent conditions are based partly on the quality of the river receiving the discharges and partly on informal industry 'norms' used by water authorities for industry in their areas.

Conditions normally limit the concentration of *suspended solids* (SS) in an effluent, the *pH* (a measure of its acidity or alkalinity) and its *biochemical oxygen demand* (BOD). The BOD is a measure of the rate at which the oxygen in a sample of water is used up by the natural self-purifying processes that break down organic pollution such as sewage or various chemicals. If the BOD is very high, all the oxygen dissolved in the water will be consumed in this purification process leaving fish and other water life starved of oxygen.

Some authorities will set limits on organic pollution by referring to the *chemical oxygen demand* (COD), a measure of the amount of oxygen used up if the organic matter is broken down by chemical reagents. Sometimes the chemical used is potassium permanganate and the corresponding limit is set in terms of the *permanganate value* (PV). Consent conditions may also limit the concentration of *metals* and other *toxic chemicals* that may be present in a discharge.

A consent that only limits the concentration of substances discharged cannot properly safeguard river quality. Any large increase in the volume of the discharge would increase the total amount of pollutant discharged, even though the concentration of the effluent remained the same. Hence, a consent should also state the maximum permitted rate of effluent discharge.

A guide to actual consent conditions applied to industrial effluents is shown in table 6.1. These standards were used as guidelines in the Yorkshire Water Authority area at the time the authority took over pollution control duties in 1974. The YWA stresses that these standards are not rigidly imposed but are modified according to the quality of water in the river and other local circumstances.[9]

During 1975, Britain came under pressure from the EEC to agree on common limits to be applied to the discharge of certain pollutants in all EEC member states. Britain rejected this proposal, maintaining that there were no consistent scientific grounds for setting uniform standards regardless of the quality of the receiving water.[10]

As a result of this disagreement, EEC member states have been permitted to decide either to follow uniform European discharge standards or to work by setting water quality objectives and controlling discharges so that these are not exceeded.

TABLE 6.1

Basis for pre-Control of Pollution Act consent conditions in the Yorkshire Water Authority Area[9]

	Type of Effluent	Processes producing such effluents	Basis for consent conditions (all units other than pH and temperature in mg/l)	
(1)	Those containing solid matter in suspension but little or no polluting matter in solution	Brick making Cement making Coal washing Power station ashponds Glass making Pottery Sand and gravel washing Water treatment plants	SS BOD pH oil	40 20 6–9 5
(2)	Those containing matter in suspension and pollutants in solution. These effluents are produced by the processing of natural organic compounds. They are similar to sewage effluent both in their effects on river quality and in their response to biological treatment (in which organic compounds are broken down by bacteria)	Food processing and manufacture Brewing Glue and gelatine manufacture Laundering Leather tanning Paper and board making Textile manufacture	SS BOD Ammoniacal nitrogen Permanganate value pH Chromium (in tannery effluents)	30 20 10 25 6–9 0·5
(3)	Those effluents in which the polluting substances are mainly in solution and are often toxic	Chemical and allied industries Engineering Gas and coke making Metal smelting and refining Plastics manufacture Plating and metal finishing Printing ink manufacture Rubber processing Soap and detergent manufacture Paint making Petroleum refining Tip drainage	Monohydric phenols Uncomplexed cyanide Total non-ferrous metals (excluding those below) Cadmium Copper Lead } none to Chromium } exceed Nickel Arsenic Zinc Total iron pH Oil Free chlorine Permanganate value BOD SS Free ammoniacal nitrogen Temperature Toxicity to fish not to exceed (level as appropriate)	1 0·1 1 0·2 0·5 2 5 6–9 5 0·1 25 30 20 10 30°C
(4)	Effluents which cause problems due to temperature variations	All cooling waters; power station cooling water discharges	Temperature not to exceed 30°C but may vary depending on volume or season	

Abbreviations: SS = suspendid solids; BOD = biochemical oxygen demand;
pH = a measure of acidity/alkalinity

RIVER QUALITY OBJECTIVES

A new system of classifying rivers, based on river quality objectives, has been advocated by the National Water Council. The suggested classification scheme is shown in table 6.2.

Using a scheme of this sort, water authorities will set river quality objectives for each river in their area. The objectives will be based on the use to which the river is put and will state the maximum concentrations of certain substances that should be present in the water; 95 per cent of all river samples taken should be within these limits, although they will not be legally enforceable.

The water authorities will then decide how much total pollution a stretch of river can accept without exceeding these limits. From this they will calculate how much each factory or sewage works may put into the river, and will then impose legally binding consent conditions.

New consents under the Control of Pollution Act will replace existing consents which, according to the National Water Council, 'are not always rationally related to environmental quality objectives—some are unnecessarily stringent, others not stringent enough'.[11] **In many cases the new consents will be more lax than those they replace. There are two reasons for this:**

(1) Some existing consents were set as 'targets to be achieved over a period of years' and not as realistic immediate standards. Many discharges were notorious for repeatedly failing to meet their consent conditions—and the old river authorities often did not attempt to enforce these standards. In 1973, 56 per cent of the total volume of sewage works effluent and 34 per cent of industrial effluent discharged into rivers in England and Wales exceeded their consent conditions and were therefore discharged illegally.[12] The review of consent conditions is expected to legitimise many of these discharges.

(2) Under the Control of Pollution Act, members of the public will for the first time be able to bring their own prosecution against a water authority or firm that exceeds its consent conditions.[13] This power of public prosecution has been introduced to deal with the conflict of interest that arose in 1974 when the new water authorities were made responsible not only for setting and enforcing standards, but also for operating sewage works—many of which have in the past exceeded their legal standards. Where a discharge could not be improved without considerable expenditure on new sewage or effluent treatment plants, the consent may be toned down to protect water authorities and others from prosecution.

Effluent discharged into a public sewer

Many firms discharge their trade effluent into public sewers

TABLE 6.2

Classification of river quality according to use, suggested by the National Water Council[11]

River Class	Quality criteria	Remarks	Current potential uses
(1A)	*Class limiting criteria* (95 percentile) (i) Dissolved oxygen saturation greater than 80% (ii) Biochemical oxygen demand not greater than 3 mg/l. (iii) Ammonia not greater than 0·4 mg/l (iv) Where the water is abstracted for drinking water, it complies with the requirements for A2** water estimates if EIFAC figures are not available (v) Non-toxic to fish in EIFAC terms (or best estimates if EIFAC figures not available)	(i) Average BOD probably not greater than 1·5 mg/l (ii) Visible evidence of pollution should be absent	(i) Water of high quality suitable for potable drinking; supply abstractions and for all other abstractions (ii) Game or other high class fisheries (iii) High amenity value
(1B)	(i) DO greater than 60% saturation (ii) BOD not greater than 5 mg/l (iii) Ammonia not greater than 0·9 mg/l (iv) Where water is abstracted for drinking water, it complies with the requirements for A2** water estimates if EIFAC terms (or best estimates if EIFAC figures not available) (v) Non-toxic to fish in EIFAC terms (or best estimates if EIFAC figures not available)	(i) Average BOD probably not greater than 2 mg/l Average ammonia probably not greater than 0·5 mg/l (iii) Visible evidence of pollution should be absent (iv) Waters of high quality which cannot be placed in class 1A because of high proportion of high quality effluent present or because of the effect of physical factors such as canalisation, low gradient or eutrophication (v) Class 1A and class 1B together are essentially the class 1 of the *River Pollution Survey*	Water of less high quality than class 1A but useable for substantially the same purposes
(2)	(i) DO greater than 40% saturation (ii) BOD not greater than 9 mg/l (iii) Where water is abstracted for drinking water, it complies with the requirements for A3** water (iv) Non-toxic to fish in EIFAC terms (or best estimates if EIFAC figures not available)	(i) Average BOD probably not greater than 5 mg/l (ii) Similar to class 2 of *RPS* (iii) Water not showing physical signs of pollution other than humic colouration and a little foaming below weirs	(i) Waters suitable for potable supply after advanced treatment (ii) Supporting reasonably good coarse fisheries (iii) Moderate amenity value
(3)	(i) DO greater than 10% saturation (ii) Not likely to be anaerobic (iii) BOD not greater than 17 mg/l*	Similar to class 3 of *RPS*	Waters which are polluted to an extent that fish are absent or only sporadically present. May be used for low grade industrial abstraction purposes. Considerable potential for further use if cleaned up
(4)	Waters which are inferior to class 3 in terms of dissolved oxygen and likely to be anaerobic at times.	Similar to class 4 of *RPS*	Waters which are grossly polluted and are likely to cause nuisance
(X)	DO greater than 10% saturation		Insignificant watercourses and ditches not usable, where objective is simply to prevent nuisance developing

Note

(a) Under extreme weather conditions (e.g. flood, drought, freeze-up), or when dominated by plant growth, or by aquatic plant decay, rivers usually in classes 1, 2 and 3 may have BODs and dissolved oxygen levels, or ammonia content outside the stated levels for those classes. When this occurs the cause should be stated along with analytical results.

(b) The BOD determinations refer to 5-day carbonaceous BOD (ATU). Ammonia figures are expressed as NH_4.

(c) In most instances the chemical classification given above will be suitable. However, the basis of the classification is restricted to a finite number of chemical determinants and there may be a few cases where the presence of a chemical substance other than those used in the classification markedly reduces the quality of the water. In such cases, the chemical classification of the water should be downgraded on the basis of the biota actually present, and the reasons stated.

(d) EIFAC (European Inland Fisheries Advisory Commission) limits should be expressed as 95% percentile limits

Abbreviations: DO = dissolved oxygen; BOD = biochemical oxygen demand;

* This may not apply if there is a high degree of reaeration.

** EEC category A2 and A3 requirements are those specified in the EEC Council Directive of 16 June 1975 concerning the Quality of Surface Water intended for Abstraction of Drinking Water in the Member States.

for treatment at sewage works. About half of all sewage works effluent in the UK now originates from industry, though this is still only a tiny proportion (about three per cent) of the total volume of industry's trade effluent.

The advantages of this policy are that:

(a) It provides an outlet for effluent from factories which are not located near a river or watercourse capable of safely receiving trade effluent.

(b) It may be more economical for industry to pay the water authority to treat its effluent than to build its own treatment plants.

(c) Mixing non-toxic industrial waste with domestic sewage helps to dilute the trade effluent to more easily treatable concentrations.

(d) Treatment processes depend on bacteria which live off—and at the same time decompose—wastes. Domestic sewage contains nutrients needed by these bacteria and is therefore a good medium in which to treat industrial wastes.

(e) Factories are charged according to the volume and strength of the effluent they discharge into sewers. This provides a financial incentive for them to conserve and recirculate water instead of discharging it with their effluent and even to extract and recover raw materials that would otherwise be lost.

Where sewage works have spare capacity, non-toxic industrial effluent can be dealt with without interfering with the treatment of sewage wastes.[14] However, in times of heavy rain, sewage works may not be able to cope with the increase in flow and may discharge sewage—and industrial effluent if it is present—into a river without full treatment, or with no treatment at all.

Another problem is that some toxic substances present in industrial effluents are 'immune' to sewage treatment processes and may actually prevent a sewage works from dealing with domestic sewage. Heavy metals, cyanides and other pollutants may inhibit or kill the bacteria used in sewage treatment.[15] Such substances may also contaminate the sludge left behind after sewage treatment, making safe disposal difficult. Toxic trade effluents may pose a threat to sewage works operators and may even, in the case of sulphuric acid or sulphur compounds, attack the concrete fabric of sewage works.

The quality of any trade effluent discharged into a public sewer therefore has to be strictly controlled. No discharge can be made without the consent of the water authority, which may impose conditions regulating the nature, composition, maximum daily quantity, and rate of the discharge and can also specify that a particular constituent of the discharge should be reduced or eliminated.[16] The consent conditions can not normally be varied for at least two years after coming into force unless some modification is needed to protect human health.¹⁷

Sources of information about consents and discharges

PUBLIC REGISTERS
You will be able to obtain full information about consents and discharges from factories and sewage works when section 41 of the Control of Pollution Act is brought into force.

Water authorities will be required to keep a register, open to the public for inspection at all reasonable times and without charge, showing details of:

(a) applications for consent to discharge effluent into a river or watercourse (but not into a public sewer);

(b) existing consents and conditions of consent for discharges to rivers or watercourses;

(c) analysis results showing the composition of effluent and river samples taken by the water authority;

(d) details of any action taken by the water authority as a result of information obtained from samples.

The water authority will be obliged to let you have a copy of any entry in the register for the payment of a small fee.

However, information about some companies' effluents may not appear on the register if the company has been able to demonstrate to the Secretary of State for the Environment that publicity would either (a) disclose valuable information about a trade secret, or (b) be contrary to the public interest.[18]

The 'public interest' exemption will probably apply only to military and defence establishments, while trade secrets are only very occasionally likely to be involved. In either case, the public register will contain no information about consents or about discharge levels, but only a statement of the grounds for keeping the information confidential.

You will also have an opportunity to comment on any new application for consent. Before the water authority decides to grant or refuse consent it must announce, for two consecutive weeks in the local newspapers, that an application has been received. However, it does not need to make an announcement if it believes the proposed discharge will have 'no appreciable effect' on the water into which it is to be made.[19]

If you do comment on or object to any application, the water authority will later be required to give you at least three weeks' advance notice if it intends to give its consent; during this period you will have the opportunity to ask the Secretary of State for the Environment if he will 'call in' the application and decide on it himself.[20]

Full details about the procedure for keeping public registers and for announcing new consent applications will be given in special regulations.

WATER QUALITY REPORTS

Until *sections 41 and 36* of the Control of Pollution Act are brought into force, water authorities will not be able to disclose information about consent conditions or discharge levels unless they have the permission of the firms involved.

Several water authorities have asked firms in their area for permission to publish data showing whether their discharges complied with the legal standards. This information has been included in special water quality reports, which also contain details about sewage effluents and the quality of river and drinking water.

Some firms have refused to allow details of their discharges to be published; you will not be able to learn anything about these discharges until the public registers of information are set up.

By June 1977, the following water authorities had published water quality reports:

Water authority	Title
Severn Trent	*Water Quality* (annual)
Welsh National Water Development Authority	*Report of the Water Quality Advisory Panel* (published each year as an appendix to the authority's annual report)
Yorkshire	*Water Quality Inheritance* (April 1974)
North West	*Water Quality Review* (annual)
Thames	*Water Quality Statistics 1973–76* (*Volume 2*)

A typical extract from one of these reports is shown in figure 6.1.

GENERAL SOURCES

A source which does not name individual firms but does contain useful general information about the quality of industrial discharges, is *Volume 2* of the *River Pollution Survey* carried out by the Department of the Environment in 1970[21] and subsequently updated.

This identifies the industries discharging into rivers in each area and gives details of the numbers and volumes of discharges, and the percentage that comply with their consent conditions (see figure 6.2). Equivalent information about discharges of sewage effluent is also given.

River catchment				**MERSEY**						
Name of company and works	Receiving stream and location of outfall (N.G.R.)	Volume Ml/d		River classification	PARAMETER	Significant quality parameters					Comments
						S.S.	B.O.D.	P.V.	Phenol	pH Formaldehyde	
Ferodo Ltd., Chapel-en-le-Frith, Stockport	Black Brook SK 056 819	Actual	0.97	Upstream 1	Consent	15	15	15	2	5·9 0·1	Manufacture of brake linings. Volume reduced investigation proceeding on treatment plant improvements.
					Mean	19	37	48	5	7·3 9·5	
		Consented	2·91	Downstream 1	Range: Maximum	27	89	112	22·4	8·8 31·2	
					Minimum	13	2·5	16	0·1	6·7 1·85	
					No. of samples	10	10	10	10	10 10	
					PARAMETER	S.S.	B.O.D.	P.V.		pH B.O.D.	
Wardle Fabrics Ltd., Whitehall Works, Chinley, Stockport	Black Brook SK 033 821	Actual	8·27	Upstream 3	Consent	40		60		5·9 40	
					Mean	120	119			7·3 212	
		Consented	2·27	Downstream 4	Range: Maximum	262	156			7·7 410	
					Minimum	64	69			6·7 110	
					No. of samples	12	12			12 12	

FIGURE 6.1

Extract from North West Water Authority's *Water Quality Review 1975*

Industry		Discharges of total effluents (which may contain some cooling water) to Non-tidal rivers		
CBI Code	Description	Total Number (% satisfactory)	Total Volume 1000s galls (% satisfactory)	Volume of Cooling Water 1000s galls
01	Brewing			
02	Brickmaking	15(93)	1769(98)	Nil
03	Cement making	2(50)	41(88)	Nil
04	Chemical and allied industries	12(50)	4938(58)	1212
05	Coal mining	87(71)	15799(61)	1040
06	Distillation of ethanol			
07	Electricity generation	21(95)	133817(100)	80300
08	Engineering	13(69)	900(49)	113
09	Food processing and manufacture	11(36)	2173(70)	1196
10	Gas and coke	9(78)	1505(93)	691
11	Glass making	2(100)	325(100)	Nil
12	Glue and gelatine			
13	General manufacturing	1(Nil)	25(Nil)	Nil
14	Iron and steel	28(57)	9258(32)	1282
15	Laundering and dry cleaning			
16	Leather tanning			
17	Metal smelting	6(67)	2101(98)	8
18	Paint making			
19	Paper and board making	3(Nil)	5940(Nil)	Nil
20	Petroleum refining			
21	Plastics manufacture			
22	Plating and metal finishing	11(36)	2172(25)	744
23	Pottery making	3(33)	125(48)	Nil
24	Printing ink etc			
25	Quarrying and mining	32(94)	8808(87)	14
26	Rubber processing	9(67)	4626(47)	1287
27	Soap and detergent			
28	Textile, cotton and man-made	1(100)	80(100)	Nil
29	Textile, wool			
30	General farming			
31	Atomic energy establishments			
50	Water treatment	10(70)	4482(91)	'150
51	Disposal tip drainage	6(83)	1489(48)	Nil
	TOTAL FOR CODES 1-51	282(71)	200373(86)	88037
52	Derelict coal mines	12	8541	
53	Other derelict mines			
54	Active coal mines	12	5740	
55	Other active mines			
	TOTAL FOR CODES 52-55	24	14281	
	TOTAL	306	214654	88037

FIGURE 6.2

Numbers and volumes of industrial effluent discharges in area of the old Trent River Authority in 1970

If you cannot get hold of specific information about a factory's discharges, a number of American sources will tell you which pollutants are likely to be discharged by different industries—see *Water Quality Criteria Databook*, Volumes 2, 3 and 5 (page 52), *Water Quality Criteria* by McKee and Woolf,[22] or *Handbook of Environmental Control*.[23] An extract from the last of these sources is shown in table 6.3. A detailed account of water pollution problems in more than 30 different industries can be found in a series of reports published by the US Environmental Protection Agency entitled *Development Documents for Effluent Limitations Guidelines and New Source Performance Standards*.[24]

TABLE 6.3
Substances present in industrial effluents[23]

Substances	Present in wastewaters from
Free chlorine	Laundries, paper mills, textile bleaching
Ammonia	Gas and coke manufacture, chemical manufacture
Fluorides	Scrubbing of flue gases, glass etching, atomic energy plants
Cyanides	Gas manufacture, plating, case hardening, metal cleaning
Sulphides	Sulphide dyeing of textiles, tanneries, gas manufacture, viscose rayon manufacture
Sulphites	Wood pulp processing, viscose film manufacture
Acids	Chemical manufacture, mines, iron and copper pickling, DDT manufacture, brewing, textiles, battery manufacture
Alkalis	Cotton and straw kiering, wool scouring, cotton mercerising, laundries
Chromium	Plating, aluminium anodising, chrome tanning
Lead	Battery manufacture, lead mines, paint manufacture, gasoline manufacture
Nickel	Plating
Cadmium	Plating
Zinc	Galvanising, zinc plating, viscose rayon manufacture, rubber processing
Copper	Copper plating, copper pickling, cuprammonium rayon manufacture
Arsenic	Sheep dipping
Sugars	Dairies, breweries, preserve manufacture, glucose and beet sugar factories, chocolate and sweet industries
Starch	Food processing, textile industries, wallpaper manufacture
Fats, oils, and grease	Wool scouring, laundries, textile industries, petroleum refineries, engineering works
Phenols	Gas and coke manufacture, synthetic resin manufacture, textile industries, tanneries, tar distilleries, chemical plants, dye manufacture, sheep dipping
Formaldehyde	Synthetic resin manufacture, penicillin manufacture
Acetic acid	Acetate rayon, pickle and beet root manufacture
Citric acid	Soft drinks and citrus fruit processing
Fluorides	Gas and coke manufacture, chemical manufacture, fertiliser plants, transistor manufacture, metal refining, ceramic plants, glass etching
Hydrocarbons	Petrochemical and rubber factories
Hydrogen peroxide	Textile bleaching, rocket motor testing
Mercaptans	Oil refining, pulp mills
Mineral acids	Chemical manufacture, mines, Fe and Cu pickling, DDT manufacture, brewing, textiles, photoengraving, battery manufacture

Predicting the impact of a discharge

Once you know the amount of effluent being discharged by a factory or sewage works, you can predict its impact on the river. This is done in two stages:

(1) First calculate the concentration of the discharged pollutants downstream of the source, after they have been diluted in the river water.

(2) Then look up the harmful effects of these concentrations on river life or river users.

To calculate the concentration of a discharged pollutant downstream of its source you need to know (a) the concentration of pollutant in the discharge and the concentration in the river upstream of the source, and (b) the flow of the effluent and the flow of the river before it receives the effluent. The flow may be given in cubic metres per second (m^3/s) or million gallons/litres per day (mgd, Ml/d). (This last figure is not as overwhelming as it sounds. A 2 foot wide channel containing 2 inches of water flowing at 4 miles per hour would carry about a million gallons per day.)

Multiplying the concentration of pollutant in the river by the flow of the river gives the *pollution load*—the weight of pollutant passing a given point in a certain period of time. The pollution load downstream of the discharge equals the load upstream plus the load contained in the discharge itself. This can be written as:

$$\left[\begin{array}{c}\text{Conc.}\\\text{downstream}\end{array} \times \begin{array}{c}\text{Flow}\\\text{downstream}\end{array}\right] = \left[\begin{array}{c}\text{Conc.}\\\text{upstream}\end{array} \times \begin{array}{c}\text{Flow}\\\text{upstream}\end{array}\right] + \left[\begin{array}{c}\text{Conc. in}\\\text{effluent}\end{array} \times \begin{array}{c}\text{Flow of}\\\text{effluent}\end{array}\right]$$

If the effluent is fully mixed in the receiving water, this equation can be rewritten as:

$$\text{Conc. downstream} = \frac{\left[\begin{array}{c}\text{Conc.}\\\text{upstream}\end{array} \times \begin{array}{c}\text{Flow}\\\text{upstream}\end{array}\right] + \left[\begin{array}{c}\text{Conc. in}\\\text{effluent}\end{array} \times \begin{array}{c}\text{Flow of}\\\text{effluent}\end{array}\right]}{[\text{flow upstream} + \text{flow of effluent}]^*}$$

Example:

A river with a cyanide concentration of 0·01 ppm (parts per million) and a flow of 600 million litres/day receives an effluent of 20 million litres/day containing 2 ppm of cyanide. What is the concentration of cyanide downstream of the discharge?

$$\text{Conc. downstream} = \frac{(0\cdot01 \times 600) + (2 \times 20)}{(600 + 20)}$$

$$= 0\cdot074 \text{ ppm}$$

In any calculation of this sort, make sure that all the figures you use are in equivalent units (in other words, if the effluent flow is given to

---

* This assumes that the river flow downstream of the source equals the flow upstream plus the flow added to it by the effluent, and that no other new flows join the river in this stretch.

you in m³/s and the river flow in mg/d, you will have to convert one to the other). Conversion factors are given on pages 175–6.

Strictly speaking, the formula for calculating the concentration of pollutants downstream of a discharge should only be used for substances that do not change either chemically or physically once they are in the river. Factors such as BOD, COD or DO (dissolved oxygen) constantly change in a river and some chemicals are transformed while they are in the water (for example, ammonia is changed to nitrite which is changed to nitrate). In these cases, the formula given above can be used only to predict the concentration of a substance in the river before any changes of this sort have had time to take place.

Assessing effluent discharges

(1) Find out from the discharger or from the water authority whether effluents are discharged into a public sewer or directly into a river.

(2) If the effluent goes directly into a river ask the water authority whether *section 41* of the Control of Pollution Act is in force:

(a) if it is, you will find details of effluent sampling results and the consent conditions on a public register;

(b) if it is not, you may find information about the discharge in the water authority's water quality report (if it produces one), provided the firm has agreed to be included in the report.

In either case you may be able to get the information directly from the discharger itself.

(3) If the effluent is discharged into a public sewer, you will probably not be able to get hold of details of sampling results unless you have the firm's permission. But you will be able to obtain sampling results showing the quality of the effluent discharged by the sewage works. In some cases, parts of this effluent (for example, metals) may be traceable to the industry or even the factory that produced them.

(4) Check that the effluent meets its consent conditions. Both the volume and the concentration of the discharge should be within the legal limits in every sample taken. However, the water authority may tolerate an increase in the concentration if the volume is correspondingly below the required level.

(5) Find out from the water authority what impact the violation of a particular standard may have on the quality of the river. The impact will vary depending on (a) the toxicity of the pollutant, (b) the nature of the river and the use to which it is put, and (c) the ability of the river to withstand increased pollution without damage.

(6) Calculate the impact of the discharge on pollution concentrations downstream. When you have done this, see if your calculations correspond to any actual changes observed in river quality (see chapter 8). You can find out how toxic the final concentrations of pollution in the river are by using the sources of information described in Part 9 of chapter 2.

7 Water Quality Objectives

Summary

There are no detailed legal standards for the quality of rivers or drinking water in the UK. River quality is often assessed by classifying rivers either according to their chemical composition or according to the variety of river life present. Water authorities have been advised to set 'river quality objectives' for their rivers based on their actual use. For some rivers this may mean that the long-term aim will no longer be to improve them to the highest standards.

River quality can also be assessed by comparing the amounts of toxic substances present to those known to be harmful to fish life. Although the EEC has proposed legally enforceable fish protection standards for rivers, the UK government maintains that these are unnecessarily stringent.

Drinking water is often taken from rivers that have been contaminated by sewage or industrial effluent. Water authorities are required by law to provide 'wholesome' drinking water and normally follow standards recommended by the World Health Organisation. Tap water in some parts of the country sometimes contains more lead or nitrates than these standards allow. Many other households receive water that fails to meet new drinking water standards that have been proposed by the EEC. Another potential source of hazard may be caused by the tiny traces of chemicals recently found in drinking water. In large concentrations, some of these substances are known to cause cancer in animals.

About 30 per cent of drinking water in the UK comes not from pure sources of water but from rivers that have been polluted with industrial and sewage effluents. During the severest drought conditions, drinking water taken from parts of the Thames actually contains more treated effluent than 'new' water and some rivers would dry up altogether without the returned effluent. River pollution can therefore have an immediate effect not only on river life but on the water supplied to our taps.

This chapter explains the standards used in the UK to ensure that river water can support fish life and provide water for public supply, and that tap water is safe to drink.

River quality

There are no legal standards in the UK for river quality, although Britain is subject to an EEC directive which lays down minimum standards for the quality of rivers used as sources of drinking water.[1]

The quality of a river can be assessed by comparing it either to a formal system of river classification or to 'criteria' which show at what concentrations of pollution river water becomes unsuitable for various uses.

Three main systems of river classification are used in the UK.

(1) Chemical classification of rivers

In a river pollution survey carried out in 1970, the Department of the Environment classified all rivers in England and Wales into four categories, based partly on the *biochemical oxygen demand* (BOD) of the water (see page 119), and partly on their ability to support fish life and the presence of toxic pollution.

A description of the four classes used is shown in table 7.1. You can see from the titles of the classes that they imply that any river below class 1 should be improved in quality. This was, in fact, the aim of the Rivers (Prevention of Pollution) Acts which were introduced for the purpose of 'maintaining or restoring the wholesomeness' of rivers.

Improving river quality is extremely expensive and requires large sums of money to be spent building new sewage and industrial effluent treatment works. In 1974, the Yorkshire Water Authority estimated that it would cost £400 million and take 15 years for it to restore all rivers in its area to class 2 quality.[2] Cuts in public spending over recent years have held back the hoped-for improvement in river quality. Between 1972 and 1973, 160 miles of river in England and Wales were upgraded to class 1. Yet in 1976, the Anglian Water Authority reported that 225 miles of river in its area alone had fallen in quality from class 1 to class 2 since 1973.[3]

The quality of river water in each of the old river authority areas (since 1974 river authorities have been replaced by regional water authorities) is shown in the Department of the Environment's *River Pollution Survey of England and Wales*.[4]

The position and quality of the rivers surveyed in 1970 is shown in a set of ten ordnance survey maps published with the first volume of the survey. Each stretch of river is coloured according to its quality class.

You can find details of subsequent changes in river quality in the annual reports of the regional water authorities, and in the special water quality reports also produced by some authorities (see page 125).

TABLE 7.1

The chemical classification of rivers used in the Department of the Environment's *River Pollution Survey*[4]

Class 1   Rivers unpolluted and recovered from pollution
(a) All lengths of rivers whatever their composition, which are known to have received no significant polluting discharges
(b) All rivers which, though receiving some pollution, have a BOD less than 3 mg/l, are well oxygenated and are known to have received no significant discharges of toxic materials or of suspended matter which affect the condition of the river bed
(c) All rivers which are generally indistinguishable biologically from those in the area known to be quite unpolluted, even though the BOD may be somewhat greater than 3 mg/l

Class 2   Rivers of doubtful quality and needing improvement
(a) Rivers not in class 1 on BOD grounds and which have a substantially reduced oxygen content at normal dry summer flows or at any other regular times
(b) Rivers, irrespective of BOD, which are known to have received significant toxic discharges which cannot be proved either to affect fish or to have been removed by natural processes
(c) Rivers which have received turbid discharges which have had an appreciable effect on the composition of the water or character of the bed but have had no great effect on the biology of the water
(d) Rivers which have been the subject of complaints which are not regarded as frivolous but which have not yet been substantiated

Class 3   Rivers of poor quality requiring improvement as a matter of some urgency
(a) Rivers not in class 4 on BOD grounds and which have a dissolved oxygen saturation, for considerable periods, below 50 per cent
(b) Rivers containing substances which are suspected of being actively toxic at times
(c) Rivers which have been changed in character by discharge of solids in suspension but which do not justify being placed in class 4
(d) Rivers which have been the subject of serious complaint accepted as well founded

Class 4   Grossly polluted rivers
(a) All rivers having a BOD of 12 mg/l or more under average conditions
(b) All rivers known to be incapable of supporting fish life
(c) All rivers which are completely deoxygenated at any time, apart from times of exceptional drought
(d) All rivers which are the source of offensive smells
(e) All rivers which have an offensive appearance, neglecting for these purposes any rivers which would be included in this class solely because of the presence of detergent foam

(2) Biological classification of rivers

The river pollution survey also uses a second method of classifying rivers, based on the variety of animal life in the river. Unpolluted and well-oxygenated water can support a good number of different fish and other species. As pollution increases, only the more resistant species survive; the variety of organisms present drops until, in very polluted rivers, only a few of the toughest species remain.

There are several different systems of classification based on this phenomenon. The Department of the Environment has used four classes in which class A, the least polluted, contained salmon, trout and a good spread of invertebrate organisms including freshwater shrimp. At the other end of the scale, in class D, rivers were fishless and barren of everything other than worms or bloodworms. The results of the biological survey can be found in the *River Pollution Survey of England and Wales*.[4]

(3) River quality objectives

As a more efficient way of deciding where to spend money on cleaning up river pollution, water authorities have been advised to set and work towards 'river quality objectives' for their rivers. The objectives will depend on the present or intended use of the river. **Instead of attempting to upgrade all rivers to class 1 of the River Pollution Survey classification, water authorities will work towards more limited, but in the short-term more realistic, goals.** Rivers required for drinking water supplies ('potable' water) will still have to be as good as class 1, but 'where neither game fishing nor water supply abstraction is likely in the foreseeable future, the quality objective might be to support coarse fish, and the implication that class 2 rivers "need improvement" would no longer apply'.[5]

To help water authorities decide on quality objectives, the National Water Council has published a new system of river classification[6]— based on, but more detailed than, the chemical classification used in the *River Pollution Survey*. The new classification is shown in table 6.2 on page 122.

Fish protection standards

River quality can also be assessed by seeing whether individual pollutants are present in concentrations high enough to damage river life.

Usually, the aim is to protect fish from pollution damage. However, since fish rely on the river for their food, pollution that damages other river life will harm them indirectly. Hence, a standard that aims to protect fish must also safeguard many other organisms.

There are at present no legal fish protection standards in the UK, but two sets of recommendations are relevant: the EIFAC water quality criteria, and the EEC river standards.

EIFAC water quality criteria

Water quality criteria for the protection of freshwater fish have been drawn up by EIFAC—the European Inland Fisheries Advisory Commission—a body sponsored by the Food and Agricultural Organisation of the United Nations.

EIFAC reports have been published for 11 different substances (references are given on page 56). Each contains a very detailed summary of the known effects on fish of exposure to different concentrations of the substance. They conclude by recommending water quality criteria which:

> should ideally permit all stages in the life cycles to be successfully completed and, in addition, should not produce conditions in a river water which would either taint the flesh of the fish or cause them to avoid a stretch of river where they would otherwise be present, or give rise to accumulation of deleterious substances in fish to such a degree that they are potentially harmful when consumed. Indirect factors like those affecting fish-food organisms must also be considered should they prove to be important.[7]

EEC river standards

The EEC has proposed that all its member states should adopt common standards for protecting freshwater fisheries. Two sets of standards have been proposed: one for freshwaters capable of supporting *salmonid* fish (game fish such as salmon, trout, char and grayling) and another for *cyprinid* fish (coarse fish such as bream, carp, chubb and roach). The draft contains maximum permitted concentrations for a number of pollutants and suggests that for other toxic substances the permitted level should not exceed a certain fraction of the lethal concentration.[8]

The Department of the Environment has opposed EEC attempts to set legally enforceable environmental standards of this sort. The Minister for the Environment has argued that:

although most rivers in the UK support thriving fish populations, many could not meet all the proposed standards ... virtually all informed opinion in the UK takes the view, and it is a view which is shared by the Government—that they are unrealistically and unnecessarily severe.[9]

Predicting damage to fish

You can predict the likely toxic effects of river pollutants on fish by consulting the sources described on pages 50–8. One of these sources (*Water Quality Criteria 1972*) also presents recommended 'safe' limits for fish, used in the United States.

The reports of pollution toxicity to fish are based on tests which measure the concentration of a substance that will kill 50 per cent of a sample of fish in the laboratory in a given time—usually 48 or 96 hours. This lethal concentration is known as the *LC50*.

The LC50 is widely used because it is based on a simple and precise test. However, like all laboratory experiments it cannot always accurately describe the effects of pollution in natural situations, because:

(a) it is based on experiments with well-fed, healthy fish and not on a natural population which would include more susceptible fish. Furthermore, fish in a river—unlike those in a tank—can swim away from polluted water;

(b) it is only concerned with pollution levels that kill fish, and not with lower concentrations that produce other harmful effects;

(c) it only describes short-term effects that occur within, say, 48 hours—but fish may be killed after a longer exposure. For example, only three per cent of trout died during a 48-hour exposure to zinc at a concentration of 40 per cent of the LC50. But if the exposure were extended to five months, 25 per cent of the fish would have died.[10]

Several detailed reviews of the methods used to test the toxicity of substances to fish have been published.[11-14]

Before using information on fish toxicity, ask the following questions.

HAS THE RIGHT SPECIES OF FISH BEEN TESTED?
Toxicity tests usually use the rainbow trout, one of the most pollution-sensitive species of freshwater fish. If trout can tolerate a certain concentration of pollutant, most other species will usually be protected. There are some exceptions. Table 7.2 shows how the LC50 of copper varies from species to species; trout may survive at concentrations that are lethal to rudd, carp and perch.[15]

TABLE 7.2
Toxicity of copper to various species of fish[15]

Species	LC50 after ten days' exposure (mg/l)*
Eel	more than 4·0
Pike	2·0
Rainbow trout	0·8
Rudd	0·4
Common carp	0·18
Perch	0·13

*Note that the higher the LC50, the more resistant the species is.

HAVE THE EFFECTS OF OTHER SUBSTANCES ON THE TOXICITY OF THE POLLUTANT BEEN TAKEN INTO ACCOUNT?

Sometimes the combined toxicity of two or more substances is greater than would be expected by adding up the individual toxicities of each substance; this phenomenon is known as 'synergism'. For example, chromium and nickel are often discharged together in effluents from the electroplating industry. At certain concentrations, a mixture of chromium and nickel is ten times more toxic to rainbow trout than would be expected if the toxicity of one metal was simply added to that of the other.[16]

Other factors such as water hardness, temperature and acidity can significantly modify the toxicity of pollutants to fish. For example, cadmium is ten times more toxic to rainbow trout in soft water than it is in hard water.[17] A drop in temperature makes zinc *less* toxic to some species of fish and *more* toxic to others.[12]

ARE FISH EXPOSED NOT TO ONE POLLUTANT BUT TO A MIXTURE OF DIFFERENT SUBSTANCES?

Most industrial effluents contain a mixture of several different substances. Sometimes these will act synergistically, but usually their joint toxicity can be predicted by adding up the sum of their individual toxicities.

For example, if a solution containing one-third of the respective lethal concentrations of each of three different pollutants is made up, the resulting mixture will have the same toxic effect as the full lethal dose of any single one of the ingredients. This principle is generally valid, but does not apply for some non-lethal effects, for substances that react together chemically, or when synergism occurs.

If the LC50 for each pollutant is taken as 1·0, then any pollutant will be present in some fraction of its own LC50. If the sum of these fractions is 1·0, the mixture will kill half of a sample of fish. A number of reports suggest that rivers where the fraction is no more than 0·2 to 0·3 of the LC50 can support fish life[18, 19] (see table 7.3). But it has also been argued that the 0·2 to 0·3 figure underestimates toxicity, especially if fish are exposed over long periods of time.[10]

TABLE 7.3
Relationship between predicted toxicity of several pollutants and the presence of fish in the River Trent[19]

Approximate range of annual mean predicted toxicity (48-hour LC50) to rainbow trout	Corresponding fish state*
0·00–0·08	Game fish
0·08–0·13	Good coarse
0·13–0·18	Fair coarse
0·18–0·24	Poor coarse
above 0·24	No fish

*Note: game fish include salmon, trout, char and grayling, which can only survive in relatively unpolluted water. Coarse fish are much hardier and include bream, carp, chubb and roach.

A report published by the US Environmental Protection Agency has recommended that the concentration of a pollutant or pollutants in a river should never exceed 0·1 of the 96-hour LC50 if fish are to be protected, and that the 24-hour average concentration should not exceed 0·05 of the LC50. A stricter limit is recommended for pollutants that build up in the food chain or in the tissues of fish: here the maximum 'safe' level is estimated as 0·05 of the 96-hour LC50, with the 24-hour average not exceeding 0·01 of the lethal concentration.[20] (The 96-hour LC50 used in America will generally represent a stricter standard—that is, a lower concentration—than the 48-hour LC50 usually used in the UK.)

Objectives to protect other river users

As well as being a source of drinking water and a home for freshwater fish, rivers are relied on by many other users. Industry uses river water to cool equipment and products and, if it is pure enough, as a raw material in the manufacture of foods and drinks; farmers use it to irrigate land and to water livestock; it serves as a means of transport, a source of power and as an important recreational facility.

All these different uses require water of different quality. Water that has been chlorinated to make it safe to drink may be poisonous to river life (hence the death of tadpoles and tiddlers kept in tap water) or unsuitable for the brewing industry. On the other hand, heavily polluted water may be perfectly suitable for keeping barges afloat.

A description of the requirements of different water users is given in *Water Quality Criteria 1972*[20] (see page 52) and in *Water Quality Criteria* by McKee and Woolf (see page 55). Water quality needs in agriculture have also been described.[21]

The EEC has introduced a directive regulating the quality of river and sea water used for human bathing[22] and plans to introduce directives covering the protection of aquatic life, water used for the breeding of shellfish, underground water sources and water used by agriculture or in industry.[23, 24] Details of the status of these proposals can be learnt from the EEC Press and Information Office (see page 188).

Drinking water standards

Chemicals entering a river used as a source of drinking water can have an immediate effect on water supplied to households. Families have bathed in pink water after a local river turned red with waste from a dye works. In another area, effluent from a sewage works took on an odour of lemons and tap water became rancid after a perfume manufacturer discharged an unusually strong effluent into a public sewer.[25]

In some instances, chemical pollution has poisoned sources of drinking water. In 1976, Northampton's water supply was threatened after cyanide waste reached the river Nene, killing 10 000 fish. Householders received fresh water from tankers and standpipes while the river recovered from the pollution believed to have been caused when an unknown tanker driver tipped 90 kg (200 lbs) of sodium cyanate into drains on an industrial estate. In the same year, Rochdale's water supply was poisoned when phenol, produced when old tyres in a quarry caught fire, reached a stream feeding a storage reservoir. More than a year after the incident, the reservoir was still out of action.

Water supplies can become contaminated by chemicals when:

(1) Waste from industry is discharged directly into rivers or into sewers which in turn discharge into a river used as a source of drinking water.

(2) Waste from industry is tipped on the land and chemicals are washed out of the tipping site by rain.
(3) Fertilisers or pesticides used on farmland drain out into rivers.
(4) Tankers carrying chemicals are involved in accidents and their loads spill out into rivers or drains.
(5) Metals from pipes in the water system (which may contain lead or zinc) dissolve into the circulating water.

There are no detailed standards for drinking water quality in UK law. The only legal obligation is contained in the 1973 Water Act which requires water authorities to supply water that is 'wholesome'. This term is usually taken to mean water that can be drunk without risk to health, and which is also clear in appearance and free from unpleasant colour, odour or taste.

Chemicals in drinking water

Water authorities in the UK frequently rely on drinking water standards recommended by the World Health Organisation in its *European Standards for Drinking Water*.[26] (Some water authorities also refer to a slightly different set of 'international standards' also published by the WHO.[27])

Drinking water standards have also been proposed by the EEC[28] whose limits are in most cases either the same as or stricter than those of the WHO. Table 7.4 shows the WHO European standards and the EEC proposed standards (although the latter may subsequently be amended).

Other useful guidelines are the American[29] and the Russian[30] standards for drinking water.

Tap water in parts of the UK sometimes fails to meet one or the other of two existing WHO European standards. Water in these areas would be judged even more unsatisfactory if the EEC's stricter standards were adopted:

NITRATES

High concentrations of nitrates in drinking water are known to cause methaemoglobinaemia in infants—a condition which can be fatal, leading to the so-called 'blue baby' deaths. The WHO recommends that drinking water should not contain more than 50 mg/l of nitrates, but states that up to 100 mg/l nitrate is tolerable. The EEC proposes a maximum single limit of 50 mg/l.

High nitrate water has been a problem in several areas. During 1975, about half of all the water supplied to households in the Anglian Water Authority area contained more than 50 mg/l of

TABLE 7.4

Existing WHO European standards[26] and proposed EEC standards[28] for chemicals in drinking water

Constituent	WHO European	Guide level	EEC Proposed Maximum admissible conc.
Arsenic	0·05		0·05
Barium	1·0		0·1
Selenium	0·01		0·01
Chromium	0·05 (Hexavalent)		0·05
Lead	0·1		0·05
Cadmium	0·01		0·005
Cyanide	0·05		0·05
PAH μg/l	0·2		0·2
Fluoride	0·9–1·7		0·7–1·5
Nitrate	50–100 (as NO_3)		50 (as NO_3)
Sulphate	250	5	250
Carbon chloroform extract	0·2–0·5		
Copper	0·05		0·050 1·5 after 16 hrs at tap
Iron	0·1	0·1	0·3
Phenols	0·001		0·0005
Manganese	0·05	0·02	0·05
Zinc	5·0		0·100 2·0 after 16 hrs at tap
Magnesium	30 if $SO_4 \geqslant 250$ 125 if $SO_4 < 250$	30	50 (MRC 5)
Hydrogen sulphide	0·05		Nil
Chloride	200–600	5	200
Anionic detergents	0·2		0·1 (lauryl sulphate)
Ammonia	0·05	0·05 (as NH_4)	0·5 (as NH_4)
Total hardness	100–500 (as $CaCO_3$)	35	(MRC 10) Hydrometric-titre
Mercury			0·001
Silver			0·01
Turbidity (SiO_2)		5	10

Constituent	WHO European	EEC Proposed Guide level	EEC Proposed Maximum admissible conc.
Turbidity (JTU)		0·1	0·3
Colour (Pt units)		5	20
Odour (dilution rate)		0	2 @ 12 °C
			3 @ 25 °C
Palatibility (dilution rate)		0	2 @ 12 °C
			3 @ 25 °C
Temperature (°C)		12	25
Conductivity		400 μS/Cm	1250μS/Cm
pH		6·5–8·5	9·5 (MRC 6·00)
Total mineral content			1500 Dry residue
Calcium		100	(MRC10)
Sodium		⩽ 20	100
Potassium		⩽ 10	12
Aluminium			0·05
Alkali level CO_3H		30	
Nitrites			0·1 (as NO_2)
Kjeldahl-nitrogen		0·05	0·5
			N+ (excluding N in NO and NO_3)
Silica			5 mg/l SiO_2 above natural level
Substances extractable in chloroform			0·1 dry residue
Dissolved oxygen			5 (as O_2)
Oxidability		1	5
			$O_2(KMnO_4)$
BOD_5		50% of initial DO	
Total organic carbon			Reason for increase in usual concentration to be given
Nickel		0·005	0·050
Phosphorus		0·3	2·0 after isolation
Antimony			0·01
Mineral oils			0·01 residue
Total pesticides μg/l			0·5
Individual pesticides μg/l			0·1
Other organo chlorine compounds μg/l			1·0

Notes
(1) All units in mg/l unless otherwise stated.
(2) The proposed EEC standards consist of (a) maximum admissible con-

nitrates, due to the contamination of water sources by nitrogen-containing fertilisers applied to farmland. The water authority has blended high nitrate water with low in order to reduce the concentration in water supplies and has been forced to stock up with nitrate-free bottled water which it will supply to mothers with newborn children if the upper WHO limit is exceeded.[31]

LEAD

The WHO recommends a maximum standard of 0·1 mg/l lead in drinking water, though it allows up to three times this amount for water that has been standing in lead pipes for more than 16 hours (or overnight). Where its upper standard is regularly exceeded, the WHO urges either that the water should be treated or that the lead piping should be replaced. A stricter standard for lead in drinking water has been proposed by the EEC which calls for a single maximum standard of 0·05 mg/l—half the existing lower WHO level.

In 1975, a survey of lead in UK drinking water revealed that some 800 000 households received water which exceeded the WHO limit of 0·1 mg/l and an estimated 1·9 million households exceeded the proposed EEC standard of 0·05 mg/l.[32]

Unknown drinking water hazards

The two examples above describe *known* hazards for which standards can and have been set. But other little-studied chemicals which may be present in drinking water may harm health without necessarily drawing attention to themselves, perhaps by slightly increasing the frequency of an already common disease. For example, it is now suspected that there is a link between 'soft' drinking water (water lacking certain mineral salts) and heart disease.[33]

Water taken from rivers carrying industrial effluents inevitably contains tiny traces of chemicals which are not affected by effluent treatment processes and which may not be detectable by normal methods of analysis. The US Environmental Protection Agency has published a study of drinking water in New Orleans which showed that tap water contained traces of 66 different organic chemicals, some of

---

centrations, (b) minimum required concentrations, and (c) guide levels. Member States would be required to set standards no greater than the maximum admissible concentrations and no lower than the minimum required concentration (MRC) where one is stipulated. Any concentration less than the guide level 'shall be considered to be entirely satisfactory'. Member States will be allowed to set 'exceptional maximum admissible concentrations' for certain substances where this is necessary because of local conditions.

which are known to cause cancer when given to laboratory animals in large doses.[34, 35] A separate study revealed a statistical relationship between drinking New Orleans water and the incidence of cancer in white males.[36] (The fact that other sectors of the New Orleans population did not show the same increased cancer rate has raised some doubts about the significance of this finding.)

Minute traces of organic compounds known to be carcinogenic in large doses have been detected in drinking water in the UK.[37] These substances, which include chlorine compounds such as chloroform, are probably formed when drinking water is disinfected with chlorine. The Water Research Centre is developing new methods of identifying these previously undetected tiny doses of chemicals ('micropollutants'). No-one has yet shown whether drinking this water may contribute to the onset of cancer in the population; however, no dose of a carcinogen has been shown to be 'safe'.

Disease-causing organisms in drinking water

Many infectious diseases are spread by water: these include *bacterial* diseases such as typhoid, cholera and dysentery, as well as those such as polio or infectious hepatitis, which are spread by *viruses*. Waterborne parasites such as worms or flukes are also responsible for illnesses.

These diseases spread when human excrement from infected persons comes in contact with sources of drinking water. The disease organisms are excreted with the faeces, but are not killed by sewage treatment and some may be able to survive in water for up to several months.

Although drinking water is often taken from rivers that have received treated sewage effluent, micro-organisms are normally first killed off by disinfection either with chlorine or with ozone. At times of critical water shortage, highly treated sewage effluent has been added to reservoirs in some parts of the world and been delivered to households as drinking water without causing any apparent ill-health.

Bacterial standards

Ideally, drinking water should be tested for all the possible disease-causing organisms that could be present. In practice, this is far too difficult and costly for routine tests, so water authorities simply look for a kind of bacteria that is normally found in water contaminated by faecal matter.

A bacteria called *E. coli*, and related bacteria in the *coliform* group, live without causing harm in the human intestine, and are always present in the faeces. Their presence in drinking water is a sign that other faecal bacteria—those that do cause disease—may also be there. The absence of these indicator bacteria normally shows that all bacteria—though not all viruses—have been removed by disinfection.

Bacterial standards for drinking water have been recommended by the WHO,[26] and by the Department of Health and Social Security.[38] The EEC's proposed directive on drinking water standards[28] also puts forward standards for bacterial quality.

These standards are in general very similar: they all prohibit the presence of *E. coli* and coliforms in water *entering* the public distribution system. They recommend that no *E. coli* should be found in water *inside* the distribution system or in water delivered to taps and they advise that, ideally, coliform bacteria should also be absent. Where coliform bacteria are found, the standards lay down the maximum number of samples, and the maximum concentration of bacteria in any sample, that should be tolerated. (Slightly laxer standards for water inside the distribution system—compared to water entering the system—are allowed, because some minor and harmless contamination caused by coliform bacteria entering the system through leaks and other sources is quite common.)

Viruses

Viruses are much more difficult to destroy than bacteria, and human diseases caused by viruses (for example, the common cold) cannot be treated by antibiotics. Viruses in drinking water are also difficult to remove and many are relatively resistant to chlorination. However, they can be destroyed if the concentration of chlorine and period of disinfection is sufficient.[39]

Experts in the 'water industry' accept that it is theoretically possible for virus diseases to be transmitted in water that has been completely freed from coliform bacteria by disinfection. Their common view is that 'the absence of cases of waterborne outbreaks of virus diseases in the United Kingdom gives reassurance that viruses at present do not pose a particular threat'.[40] It remains possible that undetected viruses in drinking water may contribute to the general level of virus infections, without necessarily causing spectacular outbreaks of severe illness.

Sources of information

You will find a description of the chemical and bacterial quality of drinking water in your area in your water authority's annual report and in its water quality report, if it produces one. These will normally describe the standards the authority follows, the number of samples of drinking water it analysed and the proportion of samples that were within the required standards.

A list of the different chemicals that have been found in water—including rivers, underground sources, effluent discharges, rainwater and tap water—has been prepared by an EEC committee. The list is drawn from reports published all over the world since 1960 and from the work of a number of research laboratories in the EEC countries and can be inspected at the Water Research Centre library[41] or obtained by writing to the Environmental Research Programme in Brussels.[42]

Using water quality objectives

(1) Find out from the water authority what river quality objective has been set for a stretch of river, and whether this objective has been met. You can check this yourself by looking at river sampling results (see chapter 9). If an objective has not been met, find out why not and what the consequences are.

(2) Enquire whether the water authority has considered setting a higher river quality objective for the stretch of river, and what the benefits and costs of achieving this would be.

(3) Use the ability of a river to support fish as an indicator of its quality. Start by finding out from the water authority or from local anglers what kinds of fish are present and how well they seem to survive. The water authority will have records of fish kills; these may indicate that shortlasting bursts of pollution—which may not have been picked up by river monitoring—have entered the river. The biological, as well as the chemical, classification of the river will give a good general idea of the river's quality.

(4) If you use recommended river quality standards as a quick guide to river pollution levels remember that there is no such thing as an absolutely 'safe' concentration of pollution (see chapter 1).

Standards involve political choices which balance up the degree of protection with the cost of greater control. People in this country will not necessarily accept the political priorities, and hence the pollution standards, that exist elsewhere. The EIFAC recommended concentrations may be acceptable in the UK, whereas the EEC proposed fish protection standards are at present disputed.

If you use a recommended fish protection standard, try and find out how much protection it is intended to give and what degree of risk it implies is acceptable.

(5) Compare the concentrations of toxic substances in the

river with those known to be hazardous to fish or other river life. Sources of information on the toxicity of river pollutants are given in chapter 2. Remember that other substances or pollutants in the river may increase the toxicity of the substance you are investigating.

(6) You will probably be able to find out about drinking water quality in your area with relatively little trouble. The results of analyses of public water supplies are shown in water authority reports, many of which also reproduce the quality standards they apply. These standards are subject to the same reservations as other quality standards (see point (4) above), so you may want to look up the hazards of particular substances for yourself using the sources described on pages 50–58.

8 River Pollution Monitoring

Summary

Concentrations of river pollution are monitored regularly by water authorities and the results of their monitoring are available to the public. Monitoring results—used in conjunction with the results of effluent sampling—will help demonstrate where pollution in the river originates and how much impact on river quality the discharges from a particular factory or sewage works are having. However, factories and sewage works are not the only causes of river pollution; other sources also have to be considered when interpreting monitoring results. Similarly, changes in river quality may be caused by variations in natural factors and not by increased or decreased discharges of effluent.

Types of monitoring

Water authorities monitor the quality both of effluents discharged into rivers and of the rivers themselves.

Effluent monitoring allows water authorities to (a) check that discharges of trade and sewage effluent comply with their consent conditions, and (b) estimate the effect of individual discharges on river pollution.

River monitoring may be used to (a) check that a river meets the water authority's river quality objective and is suitable for its present use, (b) follow the success of river pollution control policies, (c) provide advance warning of river pollution hazards that could affect drinking water or river life, and (d) locate sources of pollution and assess their impact on river quality.

Where to find river monitoring results

There are three main sources of information about pollution

in rivers: (a) the Department of the Environment's River Pollution Survey, (b) water authority reports, and (c) local river pollution studies.

A survey of the extent of river pollution in England and Wales has been published by the Department of the Environment.[1] An extract from the survey is shown in table 8.1—the classification system used is described on pages 133–4.

TABLE 8.1

Extract from *River Pollution Survey of England and Wales*, updated 1973. Comparison of the chemical quality of rivers in Great Ouse River Authority Area in 1972 and 1973

Chemical Class		Length in miles All rivers		Non-tidal rivers		Tidal rivers		Canals	
		miles	%	miles	%	miles	%	miles	%
Class 1	1972	646	68·0	607	66·6	39	100	31	100
	1973	749	78·9	710	77·9	39	100	31	100
Class 2	1972	292	30·8	292	32·1	0		0	
	1973	191	20·1	191	21·0	0		0	
Class 3	1972	8	0·8	8	0·9	0		0	
	1973	6	0·6	6	0·7	0		0	
Class 4	1972	4	0·4	4	0·4	0		0	
	1973	4	0·4	4	0·4	0		0	
Totals	1972	950		911		39		31	
	1973	950		911		39		31	

The results of water authorities' own river monitoring are summarised in their annual reports or in separate water quality reports (see page 125). More detailed results of water quality sampling can be obtained directly from the water authority, and comments on particular problems may be found in minutes of the authority's committee meetings. River monitoring results have never been considered confidential, and you should have no trouble in finding them. When *section 41* of the Control of Pollution Act is implemented (see page 124) river monitoring results will be held on special public registers kept by each water authority.

A number of river pollution research projects involve special monitoring of particular pollutants. Details of all river pollution research in the UK is shown each year in part 4 of the Department of the Environment's *Register of Research*.[2]

River maps

To interpret monitoring results, you will need to know exactly where on the river samples have been taken, whether they are upstream or downstream of effluent discharges, and how much distance there is between sampling points and discharges.

You will probably need a good river map to do this. Water authority reports will often tell you exactly where a sampling point is (for example under a certain bridge) though they may sometimes give only a coded map reference describing the exact location of the sampling point on one of the government-produced Ordnance Survey maps.[3] You can choose what scale of Ordnance Survey map you use: the largest-scale maps show the exact location of factories and sewage works in relation to the river. (One mile of land is represented by about four feet of map in the 1:1250 series of OS maps.)

A set of river maps, colouring each river according to its quality, was published in the 1975 volume of the *River Pollution Survey*,[1] though these maps do not show the position of sampling points or discharges. Some water authorities have published versions of these maps in their annual or water quality reports. This has been done by the Northumbrian, Severn Trent, Yorkshire and Anglian water authorities, though only the Northumbrian Water Authority's maps show the position of sampling points and sewage discharges. The Yorkshire Water Authority has published particularly useful maps showing the position of all industrial and sewage discharges and indicating whether these discharges were satisfactory in 1974.[4] Other water authorities publish no maps or maps with such little detail as to be effectively useless. (Wessex Water Authority's 1974–75 annual report contained a map showing in great detail the position of all major roads, motorways, motorway interchanges and motorways under construction with thin, barely visible, blue lines just showing through to indicate the position of major rivers.)

If you cannot get hold of the *River Pollution Survey* maps, and your water authority does not publish a map itself, helpful but rather small-scale river maps are available in *Waterways Atlas of the British Isles*.[5] If you need a map showing the position of public sewers one should be available for inspection at the offices of the district or borough council for the area.

Interpreting monitoring results

The rest of this chapter presents some of the key questions to ask when interpreting river monitoring results. It is directed, in particular, at readers trying to estimate the impact of discharges from a factory or sewage works on the quality of a river.

Taking all pollution sources into account

Have you taken all possible sources of pollution into account before assuming that a particular discharge is responsible?

Several different sources often discharge the same pollutant into a river. If sampling points are located in between each source, they will help you separate them out. Otherwise, you cannot assume that the pollution has come from the first source upstream of the sampling point: this is especially true for those pollutants like dissolved metals or suspended solids that are not broken down in the river.

Possible sources of pollution include not only factory and sewage works effluents but farm wastes, runoff from land near the river, overflows that relieve the pressure on sewage works during storms, rainwater from street drains, and oils or chemicals that may have been tipped directly into the river or into drains that discharge into the river.

You can find out approximately where a substance entered the river by looking at results from a series of sampling points and locating the place at which the concentration of pollutant increased. On its own this will not prove which of several sources that may discharge in this region (and between two sampling points) is responsible. Use whatever information you can find about the composition of individual discharges (see pages 124–8 to help you identify the discharger).

Bear in mind that not all discharges from industry are polluting. Over 90 per cent of the 16 000 million gallons of water discharged by industry each day is made up of cooling water that has been in contact only with heated equipment or products and is returned to the river in more or less its original condition. This water may be a little warmer than the original and contain a slightly higher concentration of substances initially present, as some of the diluting water may have been lost by evaporation. On occasions, cooling water may be cleaner when it is put back into the river than it was when it was taken out; natural-self purification processes may have had greater time to act while the water was circulating in the factory, or the firm abstracting the water may have filtered it to remove suspended solids.

Taking samples

Have samples been taken at the right points on the river for your purpose?

To find out how much pollution a factory or sewage works is causing in the river, you need to compare the quality of the river before and after the discharge, in other words upstream and downstream of the source. Locate the position of the discharge and of nearby sampling points on a river map—and make sure you know which way the river is flowing!

Find out from the water authority whether any other effluent discharges or streams or tributaries of the river join it in the stretch at which you are looking. If they meet the river between the discharge and the nearest sampling point they will make it very difficult to estimate the impact of your source. A stream containing unpolluted water will tend to dilute the concentration of pollution in the river, while a stream with polluted water, or another effluent, will increase pollution in the river. Unless the new source is separately monitored, you will not be able to tell how much of the pollution comes from it rather than from your factory or sewage works (see figure 8.1).

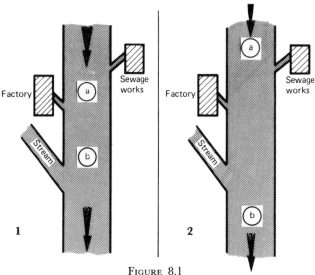

FIGURE 8.1

1. The effect of the factory's effluent on river quality can be deduced from samples taken at points (a) and (b), assuming that the effluent is fully mixed in the river before it reaches point (b).

2. The impact of the factory's discharges *cannot* be deduced from samples taken at points (a) and (b) because the change in river quality between these points is due not only to the factory's effluent but also to the inputs from the sewage works and the stream.

If there is nothing other than the effluent discharge between two sampling points, then any difference in readings taken *at the same time* at the two points *may* be due to the effluent discharge.

Analysing samples

Are river samples being analysed for the right substances?
Look up the composition of effluents being discharged into the river

and check that all the constituents are being monitored in the river itself. Remember that many trade effluents are discharged to rivers via sewage works, so you should also take sewage effluent discharges into account. If one of the constituents in an industrial discharge is not being monitored in the river, find out why not from the water authority.

Sampling results will usually tell you the maximum, minimum and average ('mean') concentration of each substance monitored. Pay special attention to the maximum concentrations, as a rare high burst of pollution can do severe damage to river life even though average concentrations may be very low.

To meet its river quality objectives, 95 per cent of all river samples analysed should be within certain limits ('95 percentile limits'). Check that the results are within these limits (shown on page 122) but do not overlook the small number of samples that are permitted to exceed them; they may still damage river life.

Changes in quality

Are changes in river quality caused by variations in natural factors?

Concentrations of pollution in the river can vary without any change occurring in the level of effluent discharges. Several natural factors can cause these changes: they include the river's own self-purifying processes, rainfall, and regular 24-hourly fluctuations in light and temperature.

Self-purification

Bacteria and other organisms that live in river water break down sewage and some organic chemicals by a process known as 'oxidation', which uses up the oxygen dissolved in the water. When organic pollution enters the river, more oxygen is needed to fuel this process and the concentration of dissolved oxygen (DO) falls. The rate at which these processes use up oxygen is called the biochemical oxygen demand (BOD). As the river purifies itself the BOD drops, and as more oxygen dissolves back into the river the DO level rises. The BOD and DO levels are commonly used as indicators of river quality; their values depend not only on the amount of pollution in the river but also on the rate at which organic matter is broken down naturally and the rate at which the river reaerates.

Certain river bacteria can also oxidise pollutants such as ammonia (which is first broken down to nitrite and then converted to nitrate) and other substances. The amount of nitrogen bound up in ammonia entering the river Tame along four stretches of its length is shown in table 8.2. The breakdown of ammonia is demonstrated by the fact that the amount of 'ammoniacal nitrogen' found in the last stretch of the river (9·22 tonnes/day) turns out to be less than the amount entering the river (11·74 tonnes/day). However, if the amount of total inorganic nitrogen (which includes not only ammonia, but also the products formed when ammonia is broken down) entering the river is examined, no decrease is found. In fact, the level of nitrogen in the river exceeds the amount entering it in discharges—probably because nitrogen from fertilisers applied to nearby land is draining into the water. The table also shows the drop in BOD down the river as self-purification takes place.

TABLE 8.2

Comparison of the weights of effluents discharged into the River Tame by sewage works and industry each day (in 1969) with the weights observed in the river (figures in tonnes/day)[6]

Subcatchment	Ammoniacal nitrogen			Total inorganic nitrogen			BOD		
	Input	Cumulative input	Observed in river	Input	Cumulative input	Observed in river	Input	Cumulative input	Observed in river
Upper	1·29	1·29	†	2·70	2·70	†	4·02	4·02	†
Upper middle	4·90	6·19	4·41	5·76	8·46	7·00	36·26	40·28	21·84
Lower middle	4·04	10·23	8·48	10·03	18·49	19·75	10·81	51·09	30·93
Lower	1·51	11·74	9·22	2·40	20·89	23·28	2·10	53·29	32·29
Loss (−) or gain (+) in system		−2·52	(21·5%)		+2·39	(11·4%)		−21·0	(39·4%)

Note: † not measured

Rainfall

Rainwater may enter a river directly or through drains and from surrounding land. A heavy storm may have several different effects on river pollution: (a) the *concentration* of industrial effluents in the river will drop because of the extra dilution (but the total *weight* of these pollutants will, of course, not change); (b) the concentration of dissolved oxygen in the river may fall because the increased volume of water is not as well aerated as the normal flow; (c) the level of suspended solids in the river will rise as particles of clay, silt and gravel are washed into the river; (d) sewage works may be overloaded by the large volumes of water reaching them from drains and be unable to treat the whole of the increased flow. Some sewage may be discharged into the river, partially treated or completely untreated, through storm overflows. The sudden arrival of this effluent can create a high BOD in the river, starving fish of oxygen. In 1976, the Welsh National Water Development Authority noted that 1000 brown trout were found dead in the River Alyn in North Wales in July 1975 after an exceptionally heavy rainstorm the previous night caused the discharge of storm sewage into the river. Concentrations of dissolved oxygen in the river had been low before the storm, and a serious deficiency in oxygen was observed afterwards.[7]

Lack of rainfall can have an equally serious impact on river quality. The volume of water reaching sewage works is reduced so the concentration of effluent they discharge is increased and has a greater effect on rivers which may already have a reduced flow. If the temperature of the water is increased because of hot weather, plants and animals in the river use up dissolved oxygen more rapidly, creating a greater oxygen demand. In severe droughts, the drop in river flow reduces the rate of reaeration and rivers may dry up altogether.

Strangely enough, the Yorkshire Water Authority reported that despite the severe drought in the summer of 1976, when river flows were the lowest in living memory, fisheries 'survived remarkably well and some actually gained considerable benefit'. The authority found that in part this was because normally overloaded sewage works were, for the first time, able to give full treatment to the reduced volume of effluent reaching them. Biological treatment processes acted more efficiently because of the increased temperature, and the end result was that the total mass of BOD added to rivers by sewage works was less than normal.[8]

Regular 24-hourly ('diurnal') changes

The regular changes in temperature and light during the day control the rate of natural processes which affect the concentration of oxygen dissolved in the river. The two processes involved are *respiration*, in which plants and animals break down food materials to obtain energy and in doing so use up oxygen, and *photosynthesis*, in which plants manufacture food substances, using sunlight, and at the same time give off oxygen. Both processes occur during daylight hours, but only respiration—which consumes oxygen—takes place at night. Hence, the concentration of dissolved oxygen in rivers tends to fall during the night (see table 8.3).

Samples of river water are usually collected during normal working hours and therefore overlook night-time conditions when dissolved oxygen is at its lowest and fish may be at greatest risk.

An important question to ask is: *have enough samples been taken to give a representative picture of river pollution?*

TABLE 8.3
The diurnal variation of dissolved oxygen in a river[9]

Time (h)	Dissolved Oxygen (mg/l)	Temperature (°C)
August 14:		
1200 noon	11·4	20·4
1400	14·1	21·0
1600	14·8	21·6
1800	13·8	21·9
2000	12·3	21·6
2200	9·7	21·3
2400	7·5	21·0
August 15:		
0200	6·3	20·4
0400	5·0	20·1
0600	4·4	19·8
0800	5·1	19·5
1000	7·1	19·5
1200	9·9	19·2

Water authorities usually sample their rivers at a frequency of between once a week to once every three months. They readily acknowledge that lack of resources prevents them from carrying out the more extensive monitoring that is often needed. According to the Severn Trent Water Authority:

> Samples, except in the case of intensive surveys, are taken during the working day, and with limitations on staff time and laboratory capability few points are sampled more frequently than fortnightly. Snap sampling is not always related to weather conditions and river flow and it is known that, when the effluent load on the river is high, the variation in quality of the river caused solely by the diurnal variation (that is, within a 24-hour period) in flow of the effluent is significant, apart from any day-to-day variation in effluent quality. Similarly, the diurnal variation in the dissolved oxygen content of the river induced by photosynthesis and respiration of rooted weed and/or algal growth can be substantial and missed completely if a single sample only is taken during the working day.[10]

Figure 8.2 shows how the concentration of copper in one river rose and fell during a 24-hour period, presumably because the strength of an industrial discharge varied during the day. Samples taken during normal working hours (shown by the short horizontal lines) failed to detect the maximum concentrations in the river, which occurred during the evening and early morning.

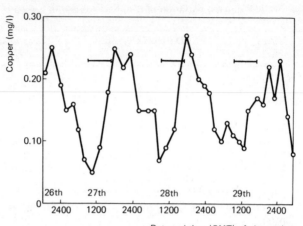

FIGURE 8.2

Changes in the concentration of copper in a river;[11] horizontal bars indicate normal sampling period

If the quality of a river or effluent varies regularly within a fixed period, at least six samples should be taken during the period in order to illustrate the kind of variation occurring.[12] If the variation depends on the time of the year, samples should be taken at least every two months; if it depends on the time of day, samples need to be taken at least every four hours.

If a river or effluent has a fairly constant strength, relatively few samples are needed to give an accurate figure for the average concentration. More samples are needed if the quality tends to fluctuate. There are special formulas for calculating the number of samples needed to give an accurate figure: for a typical river, 29 samples would be needed to estimate the average concentration of a pollutant to within 20 per cent of the true value. If the concentration in the river fluctuates widely, 55 samples would be needed. More samples would have to be taken to give a more accurate estimate—say, to within ten per cent of the real figure. In this case, 98 samples would be needed for a typical river and as many as 200 for a very variable river.[12]

Ideally, only continuous monitoring of river quality using automatic equipment can give a complete picture of changes in river pollution concentrations. Unfortunately, such equipment can usually measure only a limited number of different factors and is expensive to buy—it is therefore not used very often.

A further important question to ask is: *Could small differences in readings taken at different sites be caused by errors or the inaccuracy of the method used?*

A certain amount of inaccuracy is built in to any method of sampling and analysing water quality; on top of this, errors may be made in the way equipment is used. The way to take error and inaccuracy into account when using sampling results is explained in chapter 5 in the context of air pollution monitoring.

In 1975, the Yorkshire Water Authority found that different laboratories in its area had been producing different results when analysing identical samples. In one case, 40 per cent of analyses of a batch of identical samples were more than ten per cent out from the real value, while one in five analyses were more than 20 per cent out. These findings have been used to help standardise methods of analysis in the region.[13]

Before you conclude that small differences in concentrations measured at different sites represent real variations in river quality, try and find out from the water authority whether the differences could be accounted for by variations in the methods of analysis used or by the inaccuracy of the method itself.

Using monitoring results

(1) Locate the position of river sampling points and discharges on a map. Decide whether monitoring results taken at these positions will tell you either where on the river pollution is being discharged or what effect discharges from a particular source are having on river quality.

(2) Use river monitoring data with the results of effluent monitoring to confirm where pollution is coming from.

(3) The difference between samples taken at the same time upstream and downstream of a discharge will show you the effect of the discharge on river quality, providing that (a) the effluent has completely mixed with the river water by the time it reaches the downstream sampling point, (b) no other sources of water or effluent enter the river between the two sampling points and no water is abstracted from the river in this stretch, and (c) differences in river quality are not caused by natural factors.

(4) If you want to discover the extent to which pollution from a factory or sewage works is responsible for changes in river quality check that (a) the flow of the river has not changed between samples, (b) samples are not taken so far downstream of the discharge that changes in the BOD and other factors have been reversed by natural processes, (c) samples were taken at the same time of day to eliminate effects of diurnal variations in river quality and, (d) differences in readings could not be accounted for by the inaccuracy of the methods of analysis used.

(5) Check that you have enough samples to give a representative picture of the average river quality and to reveal short-lasting bursts of pollution. If necessary, ask the water authority to show you the full records of sampling results on which the published summaries are based. Ask whether river quality could have been worse at times of the day or night when samples were not taken.

(6) You will find more guidance on interpreting pollution monitoring results in chapter 5, which deals with air pollution monitoring; the principles apply equally to water pollution monitoring.

9 The Pollution Audit

The preceding chapters of this Handbook have attempted to show how information about toxic hazards can be found and used to answer specific problems: Where are chemicals in a river coming from? How much pollution is there in the air? How dangerous is a particular concentration of chemical?

The Handbook can also be used to answer a wider question: how effectively does a particular company protect the community, its workforce and the environment from hazards caused by the toxic substances it uses and discharges?

This, in turn, is part of a more general question: How does a company exercise power, and what are the consequences for the people who work for it, who live near it and who depend on the products it sells?

Social audits

One way in which these questions may be answered is through 'social audits': regular, independent reports on companies describing their impact on employees, the local community, the environment, and consumers.

The term 'social audit' derives from analogy with the financial audit: if there are regular reports on companies' financial performance produced for the benefit of shareholders, is it not at least as important that other sectors of society should be informed of what companies are doing to protect—or jeopardise—their interests?

The social audit would allow the public to scrutinise, understand and question any actions taken by companies that would significantly affect them. For this reason, it is essential that the reports are either produced independently of the company (as, for example, were the *Social Audit* reports listed in Appendix 5) or produced by the company and verified by an independent party having full access to the company's information.

This form of social audit should not be confused with various forms of management reports, produced for the benefit of companies themselves, and sometimes also described as 'social audits'. Essentially, these reports are designed to help managers understand the social consequences of their decisions and therefore to be able to anticipate—and even avoid—external pressure for improvement or change. These management reports are usually not published; their purpose is fundamentally different from that of the social audit advocated here, which aims to make businesses accountable to the public.

Assessing pollution

This chapter describes the basic outline of the pollution component of a social audit, the 'pollution audit'. The consumer side of a social audit has been described elsewhere, in the *Social Audit Consumer Handbook*.[1]

The pollution audit shares some common features with the system of Environmental Impact Analysis (EIA), recommended in a report produced for the Department of the Environment in 1976.[2] However, its perspective and scope are different. The EIA would be a single once-only document predicting the likely effects of a proposed new factory or other development—it would be prepared to help planning authorities decide whether or not to approve a planning application. The pollution audit would begin where the EIA leaves off—at the time when the new factory begins operating—and could be repeated at intervals so as to give a continuing picture of the quality and impact of the factory's discharges.

Questions posed by the pollution audit

The pollution audit would ask the following questions:

(1) HOW WELL ARE DISCHARGES CONTROLLED?
The audit would examine the quality of discharges from the factory and compare them firstly to the standards the company *should* meet (that is, legal limits) and secondly to standards the company *could* meet. These might include: (a) the company's own best performance in the past, (b) the quality the company may have claimed it would reach when it applied for planning permission, (c) the average performance of the whole industry, and (d) the best standards achievable by using the most effective pollution control equipment available.

(2) WHAT IMPACT DO DISCHARGES HAVE ON ENVIRONMENTAL POLLUTION CONCENTRATIONS?
The concentrations of pollution in the environment around the factory

would be studied and the audit would attempt to assess how much of it was caused by the factory's discharges.

(3) ARE DISCHARGES LIKELY TO HARM HUMAN HEALTH OR DAMAGE THE ENVIRONMENT?

The concentrations of pollution caused by the factory in the environment would be compared with those concentrations known to harm the health of human beings or other species. The audit would also report on any signs that pollution damage has already occurred, for example by scorching vegetation or killing fish, or that nuisance has been caused to the local population.

If damage has occurred or seems likely, the audit would describe the opportunities for reducing discharges and the costs and benefits that might result.

(4) IS THE COMPANY ORGANISED IN A WAY THAT WILL ALLOW IT TO COPE EFFECTIVELY WITH ENVIRONMENTAL PROBLEMS?

The audit will look for any policies that indicate the company's intentions and are specific enough to allow its performance to be compared to intent. It will ask whether the company has prearranged, effective procedures for detecting and acting on possible hazards and whether it employs trained specialist staff who fully understand the possible hazards of the company's operations and have the authority to stop hazardous practices. (The Health and Safety Executive has reported that some safety officers in industry are unable to persuade their senior managers to accept their advice unless it is accompanied by a threat of legal action from the Factory Inspectorate.[3])

(5) ARE PEOPLE WHO ARE EXPOSED TO POSSIBLE HAZARDS ADEQUATELY INFORMED?

Is the company more concerned to reassure, rather than inform, people about the possible risks of its activities? The audit will enquire whether the company has been willing to provide members of the public who want such information with details of the toxicity and quantities of pollutants it discharges.

The pollution questionnaire

The following questionnaire lists the specific questions that might be used in a pollution audit. Not all the questions listed would necessarily have to be asked—this would depend on the purpose of the enquiry and on local conditions; equally, other lines of questioning not covered in the list may be appropriate in some circumstances.

The answers to such questions will often be lengthy and complicated and not, in themselves, of interest to most readers. In such cases, the final report might carry just a summary of the relevant details supported by the 'auditor's' assessment and conclusions.

(A) Quality of pollution discharges

NUMBER AND COMPOSITION
(1) How many separate points of discharge are there and where is each located?

(2) What substances are present in each discharge?

POLLUTION CONTROLS
(3) What pollution control equipment and measures are used at each discharge point? What special procedures exist to deal with breakdowns?

(4) How efficiently was existing pollution control equipment designed to operate, and what is its current efficiency?

(5) What is the actual and expected capital expenditure and running costs of pollution control over each of the past, say, five years and each of the next five years?

MONITORING OF DISCHARGES
(6) Which constituents of each discharge are regularly monitored, and why are any others omitted?

(7) How often are samples taken? Are they taken frequently enough to give a representative picture of discharges under all conditions and especially when discharges are at their maximum?

(8) How accurate is the method of sampling and analysis?

(9) What are the results of samples that have been analysed over, say, the last year, in terms of:

(a) the average of all results;
(b) the maximum and minimum results;
(c) the concentration not exceeded in 95 per cent of all samples?

LEGAL DISCHARGE LIMITS
(10) What legal limits apply to the concentration, rate or volume of pollution discharged from the factory?

(11) In how many samples, and under what circumstances, were discharges found to exceed the legal limits?

(12) What comments, if any, have been received from the pollution control authority concerning the quality of discharges?

FUTURE ACTION

(13) What opportunity does the company have to reduce the level of discharges, in terms of:

(a) improvements that can be achieved without significant new capital expenditure;
(b) improvements that could be achieved using the best new pollution equipment and techniques available, and the cost of such measures;
(c) steps already taken, or to be taken in the near future; timetable for achieving any improvement; costs and likely results of such actions?

(B) Impact of factory discharges on environmental quality

ENVIRONMENTAL MONITORING

(14) How large is the area likely to be affected by pollution from the factory and does it contain any particularly vulnerable sectors of the population or environment (such as hospitals, newly planted forests, or rivers supplying drinking water)?

(15) At how many sites in this area are the concentrations of pollution monitored? Which pollutants discharged by the factory are monitored? Are any pollutants present in discharges not monitored, and if not, why not?

(16) Are pollution concentrations measured at points where:

(a) pollution from the factory is likely to be at its greatest;
(b) the highest concentrations of pollution from all sources in the area are likely to occur;
(c) the largest numbers of people, or the most vulnerable sectors of the population or environment, are to be found?

(17) How accurate is the method of monitoring pollution?

BACKGROUND LEVELS OF POLLUTION

(18) Are the pollutants discharged by the factory also discharged by other sources in the area, and if so, by which sources?

(19) If there are other sources of the pollutants discharged by the factory, can the factory's pollution be separated out from the 'background' pollution using available information? If so, what information is required, for example, details of the times at which different sources are inactive, details of the quality of discharges from other sources, information about wind direction or speed, rainfall, river flow etc?

(20) What are the 'background' concentrations of pollution in the area at times when the factory is not contributing to pollution (for example, because it is shut down, or, in the case of air pollution, when the wind is in the opposite direction)? Results should be stated in terms of:

(a) the average concentration;
(b) the maximum and minimum concentrations;
(c) the concentration not exceeded in 95 per cent of all samples.

The period over which results are averaged should be stated, and results given for as many different averaging times (such as three minutes, hourly, daily, annual) as are available.

FACTORY'S CONTRIBUTION TO POLLUTION

(21) What concentrations of pollution are found in the area at times when the factory is contributing to pollution? Results should be given in the same terms used in question (20).

(22) What is the best estimate of the amount of pollution caused by the factory's discharges, taking into account the possible contribution of alternative sources and the likely dispersion of pollution from all sources?

PREDICTED POLLUTION

(23) Where a new factory is involved, or where existing monitoring sites do not adequately measure pollution from the factory being studied—see question (16)—what is the best estimate of the likely pollution caused (i) by background sources, and (ii) by the factory, at points where:

(a) pollution from the factory is likely to be at its greatest;
(b) the highest concentrations of pollution from all sources in the area are likely to occur;
(c) the largest numbers of people, or the most vulnerable sectors of the population or environment, are to be found?

(24) How accurate is any estimate of the pollution likely to be caused?

(25) Under what conditions will the estimates of likely pollution made in question (23) not be valid (for example, under abnormal weather conditions, when pollution control equipment breaks down, etc.) and how often are these conditions likely to occur?

(26) When the estimates of likely pollution are not valid, what maximum concentrations of pollution may be found in the environment?

(27) Is the company able to detect conditions under which the

estimates of likely pollution will not apply, and does it have a recognised procedure for reducing discharges at these times?

(28) Are there any plans to install monitoring equipment to check on the accuracy of predictions of pollution from a new source or from a source whose impact is not already adequately monitored?

(C) Toxic effects of pollution discharges

TOXICITY DATA

(29) What are the known toxic effects resulting both from *single short-lasting exposure* and *continued long-term exposure* to different concentrations of the pollutants discharged by the factory?

The effects of the pollutants on human health and on the wellbeing of plants and animals present in the environment should be summarised. The summary should take into account possible interactions between the pollutants discharged and other pollutants or substances that may be found in the environment.

(30) Where the toxic effects of the pollutants on man and on other species actually present in the environment are not well understood, a summary should be given showing the results of toxicity tests using laboratory animals, and discussing the applicability of this information to man or other species.

QUALITY OBJECTIVES

(31) Which, if any, environmental quality objectives (recommended concentrations of toxic substances that should not be exceeded) are used or applicable in this country? How much protection are these objectives intended to give, and what possible risk may be associated with them?

LIKELY HAZARDS

(32) What are the likely hazards, to man and to other species, of exposure to (a) background concentrations of pollution from sources other than the factory, and (b) concentrations occurring when the factory is contributing to pollution?

(33) What are the likely hazards, to man and to other species, at times of maximum pollution and under the *worst possible* conditions (for example, when pollution control equipment breaks down undetected at a time when discharges are at their greatest and conditions for the dispersion of pollution are at their worst)?

REPORTED EFFECTS

(34) What damage or nuisance is known, or alleged, to have been caused by discharges from the factory?

(D) The company's internal organisation

POLICIES

(35) Does the company have any policies describing its approach to environmental issues? Are these only general statements ('do everything possible to avoid unnecessary pollution') or specific commitments against which the company's performance can be judged ('we do not operate a process for even a short time if pollution control equipment is not working at maximum efficiency')?

PERSONNEL

(36) Does the company employ specially trained staff to deal with environmental problems? Do these staff have the authority to veto potentially hazardous proposals? Are they fully aware of the potential hazards of the substances handled in and discharged from its factories?

PROCEDURES

(37) Does the company have effective procedures for detecting and responding to:

(a) reports of new hazards associated with substances used in or discharged from its factories;

(b) accidents involving the release of toxic substances into the environment;

(c) public complaint of damage or nuisance caused by discharge from its factories?

(38) Does the company regularly monitor (a) the quality and quantity of the substances it discharges, or (b) the impact of its discharges in the surrounding environment?

(E) Disclosure of information

(39) Has the company provided members of the public who might be affected by its discharges, or who request such information, with details of:

(a) possible hazards associated with substances discharged from its factories;

(b) the actual concentration or quantity of pollutants discharged;

(c) improvements requested by pollution control authorities?

Appendix 1

Published sources of information on the composition of trade-named products

Chemical dictionaries

Chemical Synonyms and Trade Names, W. Gardner and E. Cooke. (London: Technical Press Ltd., 7th Edition 1971). The only dictionary of its kind published in the UK
Hackh I.W.D. Chemical Dictionary, Julius Grant. (New York: McGraw Hill, 4th Edition 1969). Includes many British products
The Condensed Chemical Dictionary, G. G. Hawley. (New York: Van Nostrand Reinhold, 8th Edition 1971). Includes many British products
Concise Chemical and Technical Dictionary, H. Bennett. (London: Edward Arnold Ltd., 3rd Edition 1974). An American book issued by a British publisher
Handbook of Commercial Organic Chemicals, Synthetic Organic Chemical Manufacturers' Association. American Chemical Society (1966). Mainly American products, but includes an unusual section with a breakdown of trade-name products containing mixtures of chemicals
Merck Index. An Encyclopaedia of Chemicals and Drugs, (New Jersey: Merck Co. Inc., 9th Edition 1976). Includes many British products
Pharmacological and Chemical Synonyms, E. E. J. Marler. (Amsterdam: Excerpta Medica, 5th Edition 1973). Deals mainly with drugs but some information on other chemicals

Trade directories and technical books (listed by product or by industry)

Aerosols
Aerosol Review (London: Morgan Grampian). 65 pages of the brand names of aerosol products listing the ingredients and type of propellant
Adhesives
Adhesives Directory (Richmond, Surrey: A. S. O'Connor and Co. Ltd.)
Adhesives Guide, B. A. Philpott. *Design Engineering* (1968)
Adhesives Handbook, J. Shields. (London: Newnes-Butterworths, 2nd Edition 1976). Some full chemical descriptions

Chemicals and chemical industry
British Chemicals and Their Manufacturers (London: Chemical Industries Association, 1970)
Chemical Industry Directory (Tonbridge: Benn Brothers Ltd.)
European Chemical Buyers Guide (London: IPC Business Press)

Detergents
Domestic and Industrial Chemical Specialities, L. Chalmers. (London: Leonard Hill Ltd., 1966). Appendix 2 contains a full chemical description of trade products used in the manufacture of detergents and soaps
McCutcheon's Functional Materials (Glen Rock, USA: McCutcheon Publishing Co.). Mainly US products

Finishing products
Finishing Handbook and Directory (London: Sawell Publications Ltd.). UK products used in the metal, wood and plastics finishing industries

Paints
Paint Trade Manual of Raw Materials and Plant (London: Sawell Publications Ltd.). Full chemical description of products given in 'Resin Tables'
Polymers, Paint and Colour Yearbook (Redhill: Fuel and Metallurgical Journals Ltd.)

Pesticides
British Agrochemicals Association Directory (London: British Agrochemicals Assn.). Full chemical description of trade-named pesticides
Nanogen Index. A Dictionary of Pesticides and Chemical Pollutants, Compiled by Kingsley Packer. (Updated yearly). (California, USA: Nanogen International Co.). Full chemical description of trade-named pesticides
Pesticides Handbook—Entoma, Donald H. Frear. (Pennsylvania State College, USA: College Science Publishers). Full chemical description of trade-named pesticides
Pesticide Manual, H. Martin and C. R. Worthing. The British Crop Protection Council, 4th Edition 1977). Full chemical description of trade-name products and a summary of their toxic hazards

Plastics industry
British Plastics Yearbook (London: IPC Business Press)
European Plastics (Holland: Economic Documentation Office)
European Plastics Buyers Guide (London: IPC Business Press)
Laminated Plastics, D. J. Duffin. (New York: Reinhold Publishing Corp.). Products used in the US, but includes some made by British manufacturers
Modern Plastics Encyclopaedia (New York: McGraw Hill). Mainly US trade-named products; may include some made by British manufacturers
Plasticisers Guidebook and Directory (New Jersey: Noyes Data Corpora-

tion, 1972). Guide to US manufacturers with a full chemical description of trade-name products

Polymer Additives: Guidebook and Directory (New Jersey: Noyes Data Corporation, 1972). Guide to US manufacturers with a full chemical description of trade-name products

Rubber industry

British Compounding Ingredients for Rubber, B. J. Wilson. (Cambridge: W. Heffer and Sons Ltd., 1964). Full chemical description of 2000 trade-name products produced by British or Commonwealth manufacturers

New Trade Names in the Rubber and Plastics Industries (Shrewsbury: Rubber and Plastics Research Assn.). An annual publication which lists new trade names but gives very little information about their chemical composition

Rubber Chemicals, C. M. van Turnhout. (Dordrecht, Holland: D. Reidel Publishing Company). A full chemical description of trade-name products, including those manufactured in the UK

Manual for the Rubber Industry (S. Koch, Farbenfabriken Bayer AG, 1970). Gives the full chemical description of all trade-name products manufactured by Bayer

Materials and Compounding Ingredients for Rubber, J. V. Del Gatto. (New York: Rubber World Magazine). Full chemical description of trade-name products manufactured by US companies

Solvents

Handbook of Analysis of Organic Solvents, V. Sedivec and J. Flek. (Chichester: Ellis Horwood Ltd., 1976). Full chemical description of trade-name products given in Appendix 3

Solvents, Thomas H. Durrans. (London: Chapman and Hall Ltd., 8th Edition 1971). Full chemical description of trade-name products in Appendix 1

Appendix 2

Ordering US government publications

Most of the US government publications referred to in this Handbook are available either from the US Government Printing Office (if the publication has a GPO order number) or from the National Technical Information Service (if it has an NTIS order number). If a report can be obtained directly from a government agency, the address to order from is given with the reference.

Reports with a GPO number can be obtained from:

> Superintendent of Documents
> US Government Printing Office
> Washington DC 20402, USA

Reports with an NTIS number can be obtained from:

> National Technical Information Service,
> Springfield, Virginia, 22151, USA

To order a publication, write either to the GPO or the NTIS stating the appropriate title and order number and enclosing a money order in US dollars for the price of the publication. Orders from overseas should include an extra 25 per cent of the list price to cover surface mail postage. Delivery may take two to three months.

Publications can be delivered by air mail, if the additional postage cost is sent. In 1976, NTIS were charging an additional $3.00 per document for airmail delivery overseas. Prices may have increased since then.

The prices of government publications increase periodically, so it is advisable to check the latest price before ordering, either by consulting a recent list of publications or by writing to the GPO or NTIS.

Most US government publications are available in the form of microfiche cards which can be read using a special microfiche reader. If your library has such a reader, you may find that orders for microfiches cost less and are delivered more quickly. Write to the GPO or NTIS for microfiche prices or consult a recent list of their publications.

Appendix 3

Water authorities

Anglian Water Authority, Diploma House, Grammar School Walk, Huntingdon PE18 6NZ (telephone: 0480 56181)

Northumbrian Water Authority, Northumbria House, Regent Centre, Gosforth, Newcastle Upon Tyne NE3 3PX (telephone: 0632 843151)

North West Water Authority, Dawson House, Great Sankey, Warrington WA5 3LW (telephone: 092 572 4321)

Severn Trent Water Authority, Abelson House, 2297 Coventry Road, Sheldon, Birmingham B26 3PS (telephone: 021 743 4222)

Southern Water Authority, Guildbourne House, Worthing, Sussex BN11 1LD (telephone: 0903 205252)

South West Water Authority, 3–5 Barnfield Road, Exeter EX1 1RE (telephone: 0392 50861)

Thames Water Authority, New River Head, Rosebery Avenue, London EC1R 4TP (telephone: 01 837 3300)

Welsh National Water Development Authority, Cambrian Way, Brecon, Powys LD3 7HP (telephone: 0874 3181)

Wessex Water Authority, Techno House, Redcliffe Way, Bristol BS1 6NY (telephone 0272 25491/25462)

Yorkshire Water Authority, West Riding House, 67 Albion Street, Leeds LS1 5AA (telephone: 0532 448201)

River purification boards in Scotland

Clyde: City Chambers, Glasgow G2 1DU (telephone: 041 221 9600)

Forth: Colinton Dell House, West Mill Road, Edinburgh EH1 30PH (telephone: 031 441 4691)

Highland: Town House, Dingwall V15 9SD (telephone: 0349 2267)

North East: Woodside Road, Persley, Aberdeen AB2 2UQ (telephone: 0224 696647)

Solway: 39 Castle Street, Dumfries DG1 1DL (telephone: 0387 63031/2)
Tay: Council Chambers, 3 High Street, Perth (telephone: 0738 24241)
Tweed: Burnbrae, Mossilee Road, Galashiels TD1 1NF (telephone: 0896 2425)

For areas in Scotland not covered by these River Purification Boards, river pollution control is dealt with by the local authorities.

Appendix 4

Conversion factors

The following conversion factors will help you to change figures from one set of units to another. For example, from pounds per hour (lbs/h) into kilograms per second (kg/s) or from million gallons per day (mgd) into cubic metres per second (m³/s):

$$8 \text{ lbs/h} = 8 \times 0\cdot000126 \text{ kg/s} = 0\cdot001008 \text{ kg/s}$$
$$32 \text{ mgd} = 32 \times 0\cdot043813 \text{ m}^3/\text{s} = 1\cdot402016 \text{ m}^3/\text{s}$$

To convert figures in the opposite direction from that shown in the chart, for example, from kg/s into lbs/h or from m³/s into mgd, multiply them by $\dfrac{1}{\text{the conversion factor}}$, that is, divide them by the conversion factor:

$$0\cdot05 \text{ kg/s} = 0\cdot05 \times \frac{1}{0\cdot000126} = 396\cdot82539 \text{ lbs/h}$$

$$30 \text{ m}^3/\text{s} = 30 \times \frac{1}{0\cdot043813} = 684\cdot72827 \text{ mgd}$$

Multiply	by	to obtain
Weight		
pounds (lb)	0·45359	kilograms (kg)
kilograms (kg)	1000	grams (g)
grams (g)	1000	milligrams (mg)
milligrams (mg)	1000	micrograms (μg)
Length		
miles	1·6093	kilometres (km)
yards	0·9144	metres (m)
Volume		
gallons	4·546	litres (l)
cubic feet (ft³)	1728	cubic inches (in³)
cubic feet (ft³)	0·028317	cubic metres (m³)

Velocity

feet per second (ft/s)	0·3048	metres per second (m/s)
miles per hour (mph)	0·44704	metres per second (m/s)

Weight/time

pounds per hour (lb/h)	0·000126	kilograms per second (kg/s)
tons per hour (ton/h)	0·27778	kilograms per second (kg/s)

Volume/time

cubic feet per minute (ft^3/min)	0·00047195	cubic metres per second (m^3/s)
million gallons per day (mgd)	0·043813	cubic metres per second (m^3/s)

Concentration (weight/volume)

grains per cubic foot, (gr/ft^3)	0·0022884	kilograms per cubic metre (kg/m^3)
milligrams per cubic metre (mg/m^3)	1000	micrograms per cubic metre (μg/m^3)
parts per million (ppm) of a substance dissolved in water	1	milligrams per litre (mg/l)
parts per million (ppm) of a substance in air (assuming the pressure of the air is 1 atmosphere and its temperature 0°C)	$\dfrac{\text{molecular weight}}{0\cdot 0224}$	micrograms per cubic metre (μg/m^3)

Note

The *molecular weight* of a substance is the sum of the atomic weights of the atoms that make up one molecule. You can find out the numbers and kinds of atoms that make up a molecule of a substance by looking up its molecular formula in a chemical or scientific dictionary; the same dictionary will usually list the atomic weights.

For example, the molecular formula of sulphur dioxide is SO_2, showing that a molecule of sulphur dioxide is made up of one atom of sulphur (atomic weight 32) and two atoms of oxygen (atomic weight of each 16). Hence the molecular weight of SO_2 is $32 + (2 \times 16) = 64$.

Therefore,

$$1 \text{ ppm of } SO_2 = 1 \times \frac{64}{0\cdot 0224} \mu\text{g/m}^3 = 2857\cdot 14\ \mu\text{g/m}^3$$

$$500\ \mu\text{g/m}^3 \text{ of } SO_2 = 500 \times \frac{0\cdot 0224}{64} \text{ppm} = 0\cdot 175 \text{ ppm}$$

Appendix 5

Social Audit

Contents of reports 1973-1976

SOCIAL AUDIT 1
The Case for a Social Audit argues for the systematic, independent monitoring of corporate social performance. The report makes a case for direct public and consumer involvement in corporate affairs, and it looks critically at the role of shareholders who—when it comes to social issues—have taken an unearned income for unassumed responsibilities.
The Social Cost of Advertising examines advertisers' preoccupation with human motivation in buying—rather than with the qualities in a product worth selling. The report scrutinises the voluntary advertising control system and finds serious and widespread weaknesses in it. It argues for a drastic revision of standards and for complete overhaul of the control system.
The Politics of Secrecy is by James Michael, an American lawyer (ex-Nader) and an expert in freedom of information issues. Michael describes his work in Britain in trying to uncover the ultimate secret—the extent of secrecy in government. His report explains how secrecy is calculated to secure political advantage to the consistent disadvantage of Parliament, Press and the public—and it puts the case for a 'public right to know'.

SOCIAL AUDIT 2
Company Law Reform examines the existing, proposed and desirable minimum requirements for the disclosure of information by companies. The report evaluates government proposals for requiring companies to disclose more information, in the light of the government's own record on secrecy—and it identifies some 60 areas in which companies should be required to make public more information about their work.
Arms, Exports and Industry outlines the involvement of some 50 British companies in military contracting, and examines their relationship with the Government Defence Sales Organisation. The report

also describes how secrecy has been used to obstruct Parliamentary control over British arms export policies; it concludes that the case for public scrutiny of the 'defence' business, and its effect on the progress of disarmament, is overwhelming.

Something on the Press looks at the way in which several major newspapers recently handled a front-page story. The report describes some of the difficulties reporters face when trying to produce copy to tight deadlines. It describes how, in trying to get round these difficulties, many papers seem to create 'fact' from fiction; make sweeping and unwarranted assumptions about the course of events; mould the story and angle it to what are presumed to be readers' tastes; and then let the news perish, uncorrected and unfinished.

SOCIAL AUDIT 3

Tube Investments Ltd—this is a 30 000 word report on the major UK engineering group, *Tube Investments Ltd*. The report describes the performance of this company under 12 main headings: business operation; company 'philosophy'; disclosure of information; employee relations and conditions of work; minority hiring practices; race relations; health and safety at work; overseas operations; safety, quality and reliability of consumer products; military contracting; environmental responsibilities; and donations to charitable causes.

The report examines the Company's work in each of these areas—so far as was possible without co-operation from the management—and describes what was good, bad or indifferent in each case. The report also discusses some of the general problems and possibilities that might be involved in the assessment of corporate social impact, by means of 'social audits'.

SOCIAL AUDIT 4

Shareholders put to the Test looks at the theory and practice of 'shareholder democracy'. It also describes the response *Social Audit* got from a carefully selected sample of 1000 shareholders in *Tube Investments Ltd.*, when trying to table two simple resolutions on social issues, for consideration at the Company's 1974 AGM.

The Unknown Lowson Empire examines a part of the financial empire of Sir Denys Lowson and its impact on a small mining community on the Kentucky/Tennessee border in the USA.

THE ALKALI INSPECTORATE

The Alkali Inspectorate is the government agency responsible for the control of most industrial air pollution. The report on the work of this body examines the way in which it sets and enforces standards and describes the Inspectorate's relationship with local authorities and the public. The 48-page report also examines the confidential relationship between the Inspectorate and industry, and perhaps says as much about the style of government in Britain as it does about the Inspectorate itself.

SOCIAL AUDIT 5

Cable & Wireless Ltd—this report on a publicly owned corporation—describes the disastrous consequences for the Company of an irregular and unwise involvement in new and unfamiliar business. It explains how, in extricating itself, the Company succeeded in covering up losses amounting to over £2 million. The report also demonstrates how such concealments are facilitated by present auditing and accounting standards.

Advertising: the Art of the Permissible evaluates advertising standards and practice in the light of the attempts made by the industry to strengthen its voluntary control system. The report suggests that the changes that have taken place—though sweeping—remain inadequate to the needs of the present, and certainly to those of the future.

Notes in this issue briefly review seven topics relating to business and government responsibilities.

SOCIAL AUDIT 6

The **Notes** feature reviews at some length industry and government action and inaction on the question of smoking and health. It also follows up with more information on the affairs of the *Cable and Wireless* group, and calls for a public examination of the Company's affairs by the Parliamentary Select Committee on Nationalised Industries. (This Committee subsequently carried out an enquiry into the Company's work.)

Coalite and Chemical Products Ltd. is the UK's major producer of domestic solid smokeless fuels. The Company was chosen as the subject of a full Social Audit enquiry both because it plays a key role in the implementation of national clean air policy, and because it carries on operations which are potentially harmful to employees, and which cause serious environmental pollution in the neighbouring communities.

Social Audit's report on *Coalite* runs to some 20 000 words in length and concentrates on an examination of the Company's record in employee relations, health and safety at work, environmental pollution and community and consumer relations.

The report describes how the Company brought badly needed jobs to small mining communities, but at considerable cost to the local environment. It examines in detail the ironical situation whereby households near the plants that manufacture smokeless fuels should be among the last to enjoy the benefits they can bring.

SOCIAL AUDIT 7 and 8 (double issue)
Social Audit on Avon—this issue is devoted entirely to a report on *Avon Rubber Co. Ltd*. This 100 000-word report represents a unique, if not definitive, attempt to describe the major social costs and benefits of financial profitability in one of the largest public companies in the UK.

The investigation of Avon Rubber is the first ever of its kind. It was carried out by *Public Interest Research Centre* on its own initiative and at its own expense—but the research team was given extensive co-operation both by the company's management and by the trade unions concerned. It is believed that no other company in the world has ever agreed to co-operate in such an investigation on these terms.

The *Social Audit* report critically examines Avon's operations under at least 50 different headings. It reveals information which is both a credit to—and damning of—the company. But, above all, the report makes public information about Avon which virtually any other company would insist be kept secret.

However, the significance of the report on Avon goes far beyond the revelation of information about this one company's work. The report not only identifies many 'yardsticks' that might be used in assessing corporate social performance. It also focuses attention on some of the major problems and possibilities for industrial democracy in a traditional but 'progressive' British company.

The report on Avon Rubber was the last in *Social Audit's* journal series of reports: future reports will be published on a 'one-off' basis.

Social Audit reports due at the time of going to press include: 'Enquiry into the promotion of UK food and drug products in Commonwealth Developing Countries' (due Summer 1978); and 'Enquiry into the extent and effectiveness of disclosure of information to work people on toxic hazards of substances used at work' (due Spring 1979).

Information about *Social Audit* reports will be given on request. Please send a SAE to Social Audit, 9 Poland Street, London W1V 3DG, England.

References and Notes

Introduction

1. *104th Annual Report on Alkali, Etc. Works, 1967* (London: HMSO, 1968).
2. Telephone conversation 25 January 1977.
3. N. Illiff, 'The Chemical Industry and the Community', Presidential Address, Society of Chemical Industry. *Chemistry and Industry* (10 August 1968), pages 1064–77.
4. 'Social Audit on the Avon Rubber Company Ltd.', *Social Audit*, No. 7–8 (Spring 1976).
5. J. F. Gogarty, Employee Relations Planning Manager, British Leyland Cars. 'What Employees Expect to be Told'. Paper delivered to a conference, 'Presenting Financial Facts to Employees' organised by F. T. Management Services Ltd. (10 October 1975).

Health hazards and standards

1. American Conference of Governmental Industrial Hygienists. *Documentation of the Threshold Limit Values* (3rd Edition, 1971, 2nd Printing 1974).
2. G. N. Krasovskii, 'Extrapolation of Experimental Data from Animals to Man'. *Environmental Health Perspectives*, Vol. 13 (1976), pages 51–58.
3. *Principles for Evaluating Chemicals in the Environment* (US National Academy of Sciences, 1975.)
4. P. J. Lawther, 'Climate, Air Pollution and Chronic Bronchitis', *Proceedings of the Royal Society of Medicine*, Vol. 51 (1958), pages 262–64.
5. P. J. Lawther and J. A. Bonnell, 'Some Recent Trends in Pollution and Health in London—and Some Current Thoughts'. Paper

presented at the International Air Pollution Conference, Washington DC (1–11 December 1970). Reprint available from Central Electricity Generating Board, Information Services, Sudbury House, 15 Newgate Street, London EC1.
6. *Air Quality Criteria for Sulfur Oxides.* US Department of Health, Education and Welfare (January 1969). Publication number AP-50, pages 91–92.
7. H. E. Stokinger, 'Criteria and Procedures for Assessing the Toxic Responses to Industrial Chemicals', in *Permissible Levels of Toxic Substances in the Working Environment* (Geneva: International Labour Office, 1970).
8. Health and Safety at Work etc. Act 1974, *section 6 (b)*.
9. *Proposed Scheme for the Notification of the Toxic Properties of Substances*, Health and Safety Commission (London: HMSO, 1977).
10. *Air Quality Criteria for Nitrogen Oxides*, US Environmental Protection Agency (January 1971), Publication number AP-84.
11. *Sunday Times* (27 June 1976).
12. *Water Quality Criteria Data Book. Volume 2. Inorganic Chemical Pollution of Freshwater.* US Environmental Protection Agency (1971). EPA No. 18010 DPV 07/71, page 9.
13. Substances found to cause cancer in animals in any dose, however small, are prohibited for use as food additives in the US under the Food, Drug and Cosmetic Act. For a defence of this often-criticised policy, see *Cancer Prevention and the Delaney Clause*, available from the Health Research Group, 2000 P Street NW, Washington DC.
14. B. A. Bridges, 'Short Term Screening Tests for Carcinogens', *Nature*, Vol. 261 (May 20 1976), pages 195–200.
15. *Environmental Standards. A Description of United Kingdom Practice.* Department of the Environment, Pollution Paper No. 11 (London: HMSO, 1977) paragraphs 20 and 25.
16. *Threshold Limit Values for 1976*, Health and Safety Executive Guidance Note EH/15/76 (London: HMSO, 1977).
17. *Methods Used in the USSR for Establishing Biologically Safe Levels of Toxic Substances* (Geneva: World Health Organisation, 1975).
18. M. Winell, 'An International Comparison of Hygienic Standards for Chemicals in the Work Environment', *Ambio* (Royal Swedish Academy of Sciences), Vol. IV. No. 1 (1975), pages 34–36.
19. A. V. Roschin and L. A. Timofeevskaya, 'Chemical Substances in the Work Environment: Some Comparative Aspects of USSR and US Hygienic Standards', *Ambio* (Royal Swedish Academy of Sciences), Vol. IV. No. 1 (1975), pages 30–33.
20. N. F. Izmerov, *Control of Air Pollution in the USSR*, Public Health Papers No. 54 (Geneva: World Health Organisation, 1973), pages 43–44.

21. H. J. Dunster, Deputy Director-General, Health and Safety Executive. Personal communication (16 May 1977).

2 Investigating hazards

1. G. P. Henderson, *Directory of British Associations*, 5th Edition (CBD Research, 1977).
2. P. Millard, *Trade Associations and Professional Bodies in the UK*, 5th Edition (Oxford: Pergamon, 1971).
3. The British Library, Science Reference Library, *Health and Safety Literature. A Selective List of Material Held by the Science Reference Library*. (London: 1977).
4. Patrick Kinnersly, *The Hazards of Work: How to Fight Them*. (London: Pluto Press, 1973).
5. J. E. McKee and H. W. Woolf, *Water Quality Criteria*. (California State Water Quality Control Board, 1963, publication number 3-A).
6. *Hazardous Chemical Data (CHRIS)*, US Coastguard (Department of Transportation, January 1974) publication number OG-446-2.

3 Air pollution law and standards

1. *Environmental Standards. A Description of United Kingdom Practice*. Department of the Environment, Pollution Paper No. 11 (London: HMSO, 1977).
2. *Industrial Air Pollution 1975*, Health and Safety Executive (London: HMSO, 1977), page 55.
3. *Alkali Etc. Works Orders 1966 and 1971*, Available from HMSO and reprinted in *108th Annual Report on Alkali Etc. Works 1971* (London: HMSO, 1972) Appendix V. This list will be revised and reissued as regulations made under the Health and Safety at Work Act.
4. *Air Pollution Control: An Integrated Approach*, 5th Report of the Royal Commission on Environmental Pollution (London: HMSO, 1976).
5. E. A. J. Mahler, 'Standards of Emission Under the Alkali Act', Paper presented to the International Clean Air Congress, London, October 1966. Reprinted in *103rd Annual Report on Alkali Etc. Works 1966* (London: HMSO, 1967), pages 56–60.
6. *Air Pollution Control: An Integrated Approach* (see reference [4]), paragraphs 201–209.
7. *Air Pollution Control: An Integrated Approach* (see reference [4]), paragraph 216.
8. 'Notes on Best Practicable Means for Electricity Works', in

111th Annual Report on Alkali Etc. Works 1974 (London: HMSO, 1975), Appendix VII, pages 86–88.
9. *Industrial Air Pollution 1975* (see reference [2]), Appendix 2, pages 54–60.
10. This power of 'prior approval' was originally laid down by *section 9(5)* of the Alkali Etc. Works Regulation Act 1906 and has been carried on under schedule 2 of the Clean Air Enactments (Repeals and Modifications) Regulations 1974.
11. For examples, see: M. Frankel, *The Alkali Inspectorate. The Control of Industrial Air Pollution* (London: Social Audit, 1974), pages 13 and 22.
12. *The World's Air Quality Management Standards. Volume 1, The Air Quality Management Standards of the World. Volume 2, The Air Quality Management Standards of the United States* (US Environmental Protection Agency (1974), Report Number: EPA-650-9-001 A and B.
13. *Air Pollution Control: An Integrated Approach* (see reference [4]), paragraphs 122 and 124.
14. Health and Safety at Work Etc. Act 1974, *section 28(2)*.
15. Control of Pollution Act 1974, *section 79(2)*.
16. The Control of Atmospheric Pollution (Research and Publicity) Regulations 1977.
17. Control of Pollution Act 1974, *section 80(3)*. A list of the pollutants from each process about which the Alkali Inspectorate normally receives information is given in *Department of the Environment Circular 2/77* (19 January 1977).
18. Control of Pollution Act 1974, *section 81*. See also The Control of Atmospheric Pollution (Appeals) Regulations 1977.
19. Control of Pollution Act 1974, *section 81(1)(a)*.
20. Such establishments—but no others—have been exempted from disclosure under The Control of Atmospheric Pollution (Exempted Premises) Regulations 1977.
21. *Department of the Environment Circular 2/77* (19 January 1977).
22. Clean Air Act 1956 *section 3*.
23. Clean Air Act 1956, *section 6(1)* and Clean Air Act 1968, *section 3(1)*.
24. Clean Air Act 1968, *section 6*.
25. Control of Pollution Act 1974, *section 79(10)*.
26. 'Social Audit on the Avon Rubber Company Ltd.', *Social Audit*, No. 7–8 (Spring 1976).
27. Clean Air Act 1956, *section 1*. The 'permitted periods' are given in *The Dark Smoke (Permitted Periods) Regulations 1958*.
28. Clean Air Act 1956, *section 11(2)*.
29. *National Society for Clean Air Year Book*, available from NSCA, 136 North Street, Brighton (telephone: 0273 26313).

30. J. F. Garner and R. K. Crow, *Clean Air—Law and Practice*, 4th Edition (London: Shaw and Sons, 1976).
31. The charts, and instructions on their use, are shown in a series of British Standards available from the British Standards Institute, 2 Park Street, London W1. The standards involved are: BS 2742C (1957), BS 2742M (1960) and BS 2742 (1969 with addendum published in 1972).
32. The charts are reproduced in the *National Society for Clean Air Year Book* (see reference [29]).
33. The Clean Air (Emission of Grit and Dust from Furnaces) Regulations 1971.
34. Clean Air Act 1968, *section 2(4)*.
35. *Report of the Second Working Party on Grit and Dust Emissions* (London: HMSO, 1974).
36. Clean Air Act 1968, *section 7*.
37. The Oil Fuel (Sulphur Content of Gas Oil) Regulations 1976 have been issued in order to comply with Directive 75/716/EEC adopted by the Council of European Communities on 24 November 1975.
38. Public Health (Recurring Nuisances) Act 1969.
39. Public Health Act 1936, *section 99*.
40. *Report of the Working Party on the Suppression of Odours from Offensive and Selected Other Trades. Part 1, Assessment of the Problem in Great Britain* (Stevenage: Warren Spring Laboratory, 1974); *Part 2, Best Present Practice in Odour Prevention and Abatement* (Stevenage: Warren Spring Laboratory, 1975).
41. Public Health Act 1936, *section 100*.
42. Because of this, the Department of the Environment has urged local authorities to be 'fully aware of the national and local implications of any action they might be contemplating in order to reduce smell nuisance'. *Control of Smells from the Animal Waste Processing Industry, Department of the Environment Circular, 43/76* (7 May, 1976).
43. Control of Pollution Act 1974, *section 79(2)*.
44. Until 1974, the Chief Alkali Inspector's annual reports were published by HMSO for the Department of the Environment as *Annual Report on Alkali Etc. Works (year)*. From 1975 these reports have been issued by the Health and Safety Executive as *Industrial Air Pollution (year)*, available from HMSO.
45. The main source is *Compilation of Air Pollutant Emission Factors, 2nd Edition*, US Environmental Protection Agency (1973) (GPO No. EP 4.9: 42/2—see page 172 for ordering details). A number of supplements to this volume are available. Emission factors are also shown in *Handbook of Environmental Control, Volume 1* (see reference [46]).

46. R. G. Bond and C. P. Straub, *Handbook of Environmental Control, Volume 1, Air Pollution* (Cleveland, Ohio: CRC Press, 1972), page 300.
47. Source: *An Economic and Technical Appraisal of Air Pollution in the United Kingdom*, Programmes and Analysis Unit, Department of Trade and Industry and UK Atomic Energy Authority (London: HMSO, 1972).
48. D. J. Moore, 'SO_2 Concentration Measurements Near Tilbury Power Station', *Atmospheric Environment*, Vol. 1 (1967), pages 389–410.
49. 'Dispersion of Pollutants from Single Sources', in *The Investigation of Atmospheric Pollution 1958–1966*, 32nd report, Warren Spring Laboratory, Ministry of Technology (London: HMSO, 1967), pages 28–38.
50. R. S. Scorer and C. F. Barrett, 'Gaseous Pollution from Chimneys', *International Journal of Air and Water Pollution*, Vol. 6 (1962), pages 49–63.
51. Based on an illustration from: A. Coe and I. M. Coe, *Smoke Inspector's Handbook Two* (London: College of Fuel Technology, 1963).
52. *National Survey of Air Pollution 1961–71, Volume 1*, Warren Spring Laboratory, Department of Trade and Industry (London: HMSO, 1972), pages 19–20.
53. *Air Pollution Across National Boundaries: The Impact on the Environment of Sulphur in Air and Precipitation*, Royal Ministry for Foreign Affairs and Royal Ministry of Agriculture (Stockholm: Norstedt and Soner, 1971).
54. *Acid Precipitation and its Effects in Norway*, Ministry of the Environment (Oslo, 1974).
55. *Effects of Airborne Sulphur Compounds on Forests and Freshwaters*, Department of the Environment, Central Unit on Environmental Pollution, Pollution Paper No. 7 (London: HMSO, 1976).
56. In 1976, studies on the fate of sulphur dioxide emissions were being carried out by Warren Spring Laboratory and the National Environment Research Council (see *Register of Research 1976, Part IV Environmental Pollution*, Department of the Environment, 1976). Britain has also co-operated in a study on the long-range transport of sulphur compounds begun in 1972 by the Organisation for Economic Co-operation and Development.
57. A bibliography with abstracts on the subject of chimney dispersion is given in *Tall Stacks, Various Atmospheric Phenomena, and Related Aspects*, US Department of Health, Education and Welfare (May 1969), publication no. APTD 69-12, Available from NTIS (see page 172), order no. PB 194-805.
58. For example, see: P. M. Bryant, *Methods of Estimation of the*

Dispersion of Windborne Material and Data to Assist in their Application, UK Atomic Energy Authority Health and Safety Branch, AHSB (RP) R42, HL64/2719 (C10) (May 1964).
59. An assessment of the results produced by over 30 different formulas for calculating the rise of chimney gases after leaving the chimney is given in G. A. Briggs, *Plume Rise*, US Atomic Energy Commission, Office of Information Services (1969). Available from NTIS (see page 172), order no. TID-25075.
60. A. W. C. Keddie, G. H. Roberts and F. P. Williams, 'The Application of Numerical Modelling to Air Pollution in the Forth Valley', in *The Mathematical Models for Environmental Problems*, Proceedings of an International Conference held at the University of Southampton, September 8–12, 1975 (London: Pentech, 1976).
61. The Inspectorate's methods of calculating chimney heights have been described in *103rd Annual Report on Alkali Etc. Works 1966* (London: HMSO, 1967), pages 51–58, and *106th Annual Report on Alkali Etc. Works 1969* (London: HMSO, 1970), pages 56–60.
62. *Chimney Heights, Second Edition of the 1956 Clean Air Act Memorandum*, Department of the Environment (London: HMSO, 1967).
63. Based on figures given by F. E. Ireland, 'The Technical Background Leading to the Ministry's Memorandum on Chimney Heights', *Journal of the Institute of Fuel*, Vol. 36 (1963), pages 272–74.
64. G. Nonhebel, 'Chimney Design Requirements', in K. Tearle (editor), *Industrial Pollution Control: The Practical Implications* (London: Business Books, 1973), pages 158–70.
65. A. J. Clarke, D. H. Lucas and F. F. Ross, 'Tall Stacks—How Effective are They?', Paper presented to 2nd International Clean Air Conference, Washington DC (6–11 December 1970). Reprint available from Central Electricity Generating Board, Information Services, Sudbury House, 15 Newgate Street, London EC1.
66. G. Manier, 'The Errors in the Analytical Calculation of the Dispersion of Atmospheric Trace Substances (A Critical Comparison Between Measured and Calculated SO_2 Concentrations)', *Staub-Reinhalt. Luft*, Vol. 30, No. 1 (January 1970), pages 15–22.

4 Air quality objectives

1. *Environmental Standards. A Description of United Kingdom Practice*. Department of the Environment, Pollution Paper No. 11 (London: HMSO, 1977), page 25.

2. European Parliament, Working Document 399/75 (5 December 1975).
3. *Air Quality Criteria and Guides for Urban Air Pollutants. Report of a WHO Expert Committee*, World Health Organisation, Technical Report Series, No. 506 (Geneva: WHO 1972).
4. S. Radcliffe, 'Local Authorities and the Monitoring and Assessment of Air Quality', *Greater London Intelligence Quarterly*, No. 33 (December 1975), pages 11–15.
5. *Winter Mean Concentration of Smoke 1972–3* and *Winter Mean Concentration of Sulphur Dioxide 1972–3*. Both maps are available from Department of the Environment Map Library, Prince Consort House, 27–29 Albert Embankment, London SE1.
6. See B. D. Gooriah, A. W. C. Keddie and F. P. Williams, 'Smoke and SO_2 Contour Maps of the UK', Warren Spring Laboratory, Investigation of Air Pollution, Standing Conference of Cooperating Bodies, paper no. SCCB 85/4 (1 December 1975).
7. *Report of a Collaborative Study on Certain Elements in Air, Soil, Plants, Animals and Humans in the Swansea–Neath–Port Talbot Area Together with a Report on a Moss Bag Study of Atmospheric Pollution Across South Wales* (Welsh Office, 1975), page 211.
8. *Air Pollution Control: An Integrated Approach*, 5th Report of the Royal Commission on Environmental Pollution (London: HMSO, 1976), paragraph 167.
9. *Air Pollution Control: An Integrated Approach* (see reference [8]), paragraph 177.
10. *Air Pollution Control: An Integrated Approach* (see reference [8]), paragraphs 200 and 215–18.
11. Commission of the European Communities, 'Environment Programme 1977–1981', *Bulletin of the European Communities*, Supplement 6/76.
12. Commission of the European Communities, 'Proposal for a Council Directive (EEC) on Air Quality Standards for Lead', COM (75) 166 (16 April 1975).
13. Commission of the European Communities, 'Proposal for a Council Directive Concerning Health Protection Standards for Sulphur Dioxide and Suspended Particulate Matter in Urban Atmospheres', COM (76) 48 final (19 February, 1976).
14. Second Consultation Paper on the Royal Commission on Environmental Pollution's 5th Report, Department of the Environment (3 December 1976).
15. EEC Press and Information Office, 20 Kensington Palace Gardens, London W8 (telephone: 01 727 8090).
16. *Environmental Quality 1976*, 7th Annual Report of the Council on Environmental Quality (September 1976), available from the

US Government Printing Office—see page 172—stock number 041-010-00031-2.
17. See, *Air Pollution Considerations in Residential Planning, Volume 1, Manual*, US Environmental Protection Agency (July 1974), Publication No. EPA-450/3-74-046a.
18. *Washington Post* (12 November 1976).
19. N. F. Izmerov, *Control of Air Pollution in the USSR*, Public Health Papers No. 54 (Geneva: World Health Organisation, 1973), pages 129–31.
20. *Control of Air Pollution in the USSR* (see reference [19]) pages 43–44.
21. *The World's Air Quality Management Standards, Volume 1, The Air Quality Management Standards of the World. Volume 2, The Air Quality Management Standards of the United States*, US Environmental Protection Agency (1974), Report Number: EPA-650-9-001 A and B.
22. *Threshold Limit Values for 1976*, Health and Safety Executive, Guidance Note EH/15/76.

5 Air pollution monitoring

1. M. P. M. Weatherley, B. D. Gooriah and J. Charnock, *Fuel Consumption, Smoke and Sulphur Dioxide Emissions and Concentrations, and Grit and Dust Deposition in the UK, up to 1973–4*, Warren Spring Laboratory (1976), Report No. LR 214 (AP).
2. See *Report of a Collaborative Study on Certain Elements in Air, Soil, Plants, Animals and Humans in the Swansea–Neath–Port Talbot Area Together with a Report on a Moss Bag Study of Atmospheric Pollution Across South Wales* (Welsh Office, 1975).
3. *The Monitoring of the Environment in the United Kingdom*, Central Unit on Environmental Pollution, Department of the Environment (London: HMSO, 1974).
4. H. Steward and R. Derwent, 'Preliminary Findings of the Five Towns Survey', paper given at the 41st Clean Air Conference, Cardiff. National Society for Clean Air (October 1974).
5. G. McInnes, 'Multi-element and Sulphate in Particulate Surveys: Monitoring Locations, Sampling and Analytical Methods, and Preliminary Reporting System' (Warren Spring Laboratory, 1977).
6. *The Investigation of Air Pollution, Directory of Sites used from the Beginning of the Co-operative Investigation up to (date). Part 1, Daily Observations of Smoke and Sulphur Dioxide*, Warren Spring Laboratory. (These and other reports from this laboratory are available from PO Box 20, Gunnels Wood Road, Stevenage, Herts (telephone: 0438 3388) or from HMSO.)

7. *The Investigation of Air Pollution, National Survey of Smoke and Sulphur Dioxide* (date) Warren Spring Laboratory.
8. *National Survey of Air Pollution 1961–71*, Warren Spring Laboratory: Volume 1, General Introduction, United Kingdom—a summary, S.E. Region, Greater London area (London: HMSO, 1972), £3.50. Volume 2, S.W. Region, N.W. Region, Wales (London: HMSO, 1972), £3.50. Volume 3, E. Anglia, E. Midlands, W. Midlands (London: HMSO, 1973), £3.50. Volume 4, Yorkshire and Humberside, Northern (London: HMSO, 1976), £12.00. Volume 5, Scotland and Northern Ireland (London: HMSO, 1976), £8.00.
9. *Winter Mean Concentration of Smoke 1972–3* and *Winter Mean Concentration of Sulphur Dioxide 1972–3*. Both maps available from Department of the Environment Map Library, Prince Consort House, 27–29 Albert Embankment, London SE1.
10. *The Investigation of Air Pollution, Deposit Gauge and Lead Dioxide, Monthly and Seasonal Results* (date), Warren Spring Laboratory.
11. *Register of Research, Part IV Environmental Pollution* (annual) from Department of the Environment, Headquarters Library, 2 Marsham Street, London SW1.
12. *National Society for Clean Air Year Book* (annual), from NSCA, 136 North Street, Brighton (telephone: 0273 26313).
13. The CERL monitor is described in *Atmospheric Environment*, Vol. 1 (1967), pages 619–36.
14. A full description of the classes of districts used in the National Survey is given in the *Directory of Sites* (reference[6]) and in the tables of results (reference[7]).
15. F. P. Williams, 'National Survey of Smoke and Sulphur Dioxide, North West Region', Paper presented to Standing Conference of Co-operating Bodies, The Investigation of Air Pollution, Warren Spring Laboratory (8 June 1970). Paper No. SCCB 74/8.
16. Doncaster and District Air Pollution Survey, 1st report, 1971–72, Air Pollution Research Unit, Department of Geography, University of Sheffield (August 1972).
17. A. Garnett, Evidence at planning inquiry into application by Coalite and Chemical Products Ltd., held at Doncaster College of Technology (3–27 February 1970).
18. C. E. Zimmer and R. I. Larsen, 'Calculating Air Quality and Its Control', *Journal of the Air Pollution Control Association*, Vol. 15 (1965) No. 12, pages 565–72.
19. *National Survey of Air Pollution 1961–71, Volume 1*, Warren Spring Laboratory (London: HMSO, 1972), pages 5–6.
20. M. P. M. Weatherley, 'Interpretation of Data from Air Pollution Surveys in Towns, Taking into Account the Siting of the Instruments', *International Journal of Air and Water Pollution*, Vol. 7 (1963), pages 981–87.

21. *Instruction Manual, National Survey of Smoke and Sulphur Dioxide*, Warren Spring Laboratory, Ministry of Technology (1966).
22. The accuracy, and sources of possible errors, of the National Survey results are discussed in *National Survey of Air Pollution 1961–71, Volume 5* (see reference [8]), pages 112–18.
23. P. A. Read, 'The National Survey: Some Problems of Instrumental Accuracy and Maintenance', Paper presented to the Standing Conference of Co-operating Bodies, The Investigation of Air Pollution, Warren Spring Laboratory (May 1971).
24. Paper presented to the Standing Conference of Co-operating Bodies, The Investigation of Air Pollution, Warren Spring Laboratory, SCCB 81/3 (12 November 1973).
25. Evidence given at planning inquiry into application by Coalite and Chemical Products Ltd., held at Doncaster College of Technology (3–27 February 1970).

6 River pollution law and standards

1. For a non-legalistic guide to pollution law, see: J. McLoughlin, *The Law and Practice Relating to Pollution Control in the United Kingdom* (London: Graham and Trotman, 1976). A much more detailed guide is given in: J. F. Garner, *Control of Pollution Encyclopaedia* (London: Butterworth and Co., 1977). A guide to the opportunities for public participation opened up by the Control of Pollution Act is given in: B. Zaba and R. Macrory, *Control of Pollution Campaign Manual* (London: Friends of the Earth, 1977), available from FOE, 9 Poland Street, London W1. Another useful reference book is: A. S. Wisdom, *The Law of Rivers and Watercourses*, 3rd Edition (London: Shaw and Sons, 1976).
2. *Who's Who in the Water Industry*, An Official Publication for the National Water Council (Watford: Wheatland Journals, annual).
3. Control of Pollution Act 1974, *section 31*.
4. Control of Pollution Act 1974, *section 34(4)*.
5. Control of Pollution Act 1974, *sections 37(1) and 38(1)*.
6. Control of Pollution Act 1974, *section 38(3)(a)*.
7. Control of Pollution Act 1974, *section 46(1)*.
8. Control of Pollution Act 1974, *section 38(4)*.
9. Yorkshire Water Authority, *Water Quality Inheritance 1st April, 1974*, pages 62–65.
10. *EEC Draft Decision on Dangerous Substances in the Aquatic Environment: The Scientific Issues*, Note by an Interdepartmental Group of Experts, Department of the Environment (1975).
11. 'Review of Discharge Consent Conditions—Consultation Paper', National Water Council (February 1977).

12. *River Pollution Survey of England and Wales, Updated 1973, River Quality and Discharges of Sewage and Industrial Effluents*, Department of the Environment (London: HMSO, 1975), page 3.
13. The Control of Pollution Act repeals *section 11* of the Rivers (Prevention of Pollution) Act 1961 which restricted the power of prosecution to river authorities and the Attorney General.
14. A. B. Wheatland, M. G. W. Bell and A. Atkinson, 'Pilot Plant Experiments on the Effects of Some Constituents of Industrial Waste Waters on Sewage Treatment', *Journal of the Institute of Water Pollution Control*, No. 6 (1971). Available from the Water Research Centre as Reprint No. 653.
15. A. G. Boon and B. J. Borne, 'Minimising the Problem of Treatment and Disposal of Industrial Effluent', Paper presented to Conference on Preventing Industrial Pollution (June 1971). Available from the Water Research Centre as Reprint No. 645.
16. Trade effluent discharges to public sewers are controlled under the Public Health (Drainage of Trade Premises) Act 1937 and the Public Health Act 1961.
17. Control of Pollution Act 1974, *section 45*.
18. Control of Pollution Act 1974, *section 42(1)*.
19. Control of Pollution Act 1974, *section 36*.
20. Control of Pollution Act 1974, *section 36(6)*.
21. *Report of a River Pollution Survey of England and Wales 1970, Volume 2, Discharges and Forecasts of Improvement*, Department of the Environment (London: HMSO, 1972). An updating volume for 1973 has been published (see reference [12]), and a further updating volume for 1975 was published in 1978.
22. J. E. McKee and H. W. Woolf, *Water Quality Criteria*, California State Water Quality Control Board (1963), publication number 3-A.
23. R. C. Bond and C. P. Straub, *Handbook of Environmental Control, Volume 4, Wastewater Treatment* (Cleveland, Ohio: CRC Press, 1974).
24. *Development Documents for Effluent Limitations Guidelines and New Source Performance Standards* (US Environmental Protection Agency) have been published for the following industries. The GPO order number (see page 172) follows the title of each report:

Industry	GPO order number
Pulp and Paper—*Unbleached Kraft and Semichemical Pulp*	5501-00910
Builders Paper and Roofing Felt	5501-00909
Red Meat Processing	5501-00843
Dairy Product Processing	5501-00898
Grain Processing	5501-00844

Citrus, Apple and Potato Processing	5501-00790
Catfish, Crab, Shrimp and Tuna Processing	5501-00920
Beet Sugar	5500-00117
Cane Sugar Refining	5501-00826
Textile Mills	5501-00903
Cement Manufacturing	5501-00866
Feedlots	5501-00842
Electroplating—Copper, Nickel, Chrome and Zinc	5501-00816
Organic Chemicals—Major Organic Products	5501-00812
Inorganic Chemicals—Major Inorganic Products	5502-00121
Plastics—Synthetic Resins	5501-00815
Soap and Detergent Manufacturing	5501-00867
Basic Fertiliser Chemicals	5501-00868
Petroleum Refining	5501-00912
Steel Making	5501-00906
Bauxite Refining	5500-00118
Primary Aluminium Smelting	5501-00817
Secondary Aluminium Smelting	5501-00819
Phosphorous Derived Chemicals	5503-00078
Steam Electric Powerplants	n.a.
Power Plants—Cooling Water Intake Structure Technology	n.a.
Ferroalloys—Smelting and Slag Processing	5501-00780
Leather Tanning and Finishing	5501-00818
Insulation Fibreglass	5501-00781
Flat Glass	5501-00814
Asbestos—Building, Construction and Paper	5501-00827
Rubber Processing—Tyre and Synthetic	5501-00885
Timber Products—Plywood, Hardboard and Wood Preserving	5501-00853
	EPA order number
Meat Rendering	EPA 440/1-74/031D
Formulated Fertiliser	EPA 440/1-74/042A
Asbestos—Textiles, Friction Material and Seeding Devices	EPA 440/1-74/035A
Grain Mills—Animal Feeds, Breakfast, Cereals and Wheat Starch	EPA 440/1-74/039A
Non-ferrous Metals Manufacturing—Copper, Lead and Zinc	EPA 440/1-75/032

7 Water quality objectives

1. 'Council Directive of 16 June 1975 Concerning the Quality

Required of Surface Water Intended for the Abstraction of Drinking Water in the Member States' (75/440/EEC), *Official Journal of the European Communities* No. L 194/26 (25 July 1975).
2. Yorkshire Water Authority, *Water Quality Inheritance on 1 April 1974*, page 83.
3. *Municipal Engineering* (7 January 1977).
4. *Report of a River Pollution Survey of England and Wales 1970, Volume 1*, Department of the Environment (London: HMSO, 1971), and *River Pollution Survey of England and Wales Updated 1973, River Quality and Discharges of Sewage and Industrial Effluents*, Department of the Environment (London: HMSO, 1975). A similar updating volume for 1975 was published in 1978.
5. D. Walker (Assistant Director-General, National Water Council), 'Water Pollution Control—the Priorities', *Water* (May 1975).
6. 'Review of Discharge Consent Conditions—Consultation Paper', National Water Council (February 1977).
7. 'Report on Cadmium and Freshwater Fish', European Inland Fisheries Advisory Commission, Technical Paper No. 30, Food and Agriculture Organisation of the United Nations (Rome, 1977).
8. 'Proposal for a Council Directive on the Quality Requirements for Waters Capable of Supporting Freshwater Fish', *Official Journal of the European Communities*, No. C202 (28 August 1976).
9. Department of the Environment Press Notice 199 (25 April 1977).
10. V. M. Brown, D. G. Shurben and D. Shaw, 'Studies on Water Quality and the Absence of Fish from Some Polluted English Rivers', *Water Research* Vol. 4, pages 363–82. Available from the Water Research Centre as Reprint No. 606.
11. J. B. Sprague, 'Measurement of Pollutant Toxicity to Fish. 1. Bioassay Methods for Acute Toxicity', *Water Research* Vol. 3 (1969), pages 793–821.
12. J. B. Sprague, 'Measurement of Pollutant Toxicity to Fish. 2. Utilising and Applying Bioassay Results', *Water Research* Vol. 4 (1970) pages 3–32.
13. J. B. Sprague, 'Measurement of Pollutant Toxicity to Fish. 3. Sublethal Effects and "Safe" Concentrations', *Water Research* Vol. 5 (1971), pages 245–66.
14. R. Lloyd, 'Problems in Determining Water Quality Criteria for Freshwater Fisheries', *Proceedings of the Royal Society (Series B. Biological Sciences)*, Vol. 180 (1972), pages 439–49.
15. 'Fish and Water-Quality Criteria', Notes on Water Pollution No. 65, Water Research Centre.
16. 'Nickel and Chromium—Synergistic Toxicity to Rainbow Trout', Water Research Centre, Medmenham Laboratory. Note distributed during Open Days (4–6 May 1977).
17. A. L. Downing and R. D. Edwards, 'Effluent Standards and the

Assessment of the Effects of Pollution on Rivers', in: 'Symposium on Effluent Standards', *Water Pollution Control*, Vol. 68 (1969) No. 3, pages 283–99.
18. J. F. de L. G. Solbe, 'The Relationship Between Water Quality and the Status of Fish Populations in Willow Brook', *Water Treatment and Examination*, Vol. 22 (1973), pages 41–61. Available from the Water Research Centre as Reprint No. 702.
19. D. J. Brewin, M. S. T. Chang, K. S. Porter and A. E. Warn, 'Trent Mathematical Model: Development', Paper presented at Symposium on Advanced Techniques in River Basin Management: The Trent Model Research Programme, held at University of Birmingham (11–14 July 1972). Available from Water Research Centre as Reprint No. 678.
20. *Water Quality Criteria 1972*, US Environmental Protection Agency (March 1973), EPA No. R3-73-033, page 123.
21. J. H. Williams, 'Water Quality Criteria for Crop Irrigation (Chloride, Boron and Sodium)', *Agricultural Development and Advisory Service Review* No. 7 (Winter 1972), pages 106–22, published by HMSO.
22. Council Directive Relating to Pollution of Sea Water and Fresh Water for Bathing, adopted on 8 December 1975. *Official Journal of the European Communities*, No. L 31 (5 February 1976).
23. Commission of the European Communities, 'Communication from the Commission to the Council Concerning the State of Progress of the European Community's Environment Programme as at 15 November 1976', COM (76) 639 final.
24. *Euroforum* 41/76 (16 November 1976).
25. H. Fish, 'Some Cases of Quality Affecting Water Use', *Effluent and Water Treatment Journal* (October 1973), pages 629–36.
26. *European Standards for Drinking Water*, 2nd Edition (Geneva: World Health Organisation, 1970).
27. *International Standards for Drinking Water*, 3rd Edition (Geneva: World Health Organisation, 1971).
28. 'Proposal for a Council Directive Relating to the Quality of Water for Human Consumption', *Official Journal of the European Communities*, No. C 214/2 (18 September 1975).
29. 'National Interim Primary Drinking Water Standards', US Environmental Protection Agency, 1975 *Federal Register* 40, No. 248 59566. In: D. G. Miller, 'Research Aspects: Water', *Symposium on Developments in Water and Sewage Treatment*, Institution of Water Engineers and Scientists (London: IWES, 1976).
30. 'All Union State Standards 2874–73', State Committee of Standards of USSR, Council of Ministers. In: D. G. Miller, 'Research Aspects: Water', *Symposium on Developments in Water and Sewage*

Treatment, Institution of Water Engineers and Scientists (London: IWES, 1976).
31. Anglian Water Authority, Annual Report and Accounts 1975–76, pages 35–36.
32. *Lead in Drinking Water, A Survey in Great Britain, 1975–1976*, Report of an Interdepartmental Working Group, Department of the Environment, Pollution Paper No. 12 (London: HMSO, 1977).
33. Water Research Centre, Annual Report 1975–76, pages 38–39.
34. *New Orleans Area Water Supply Study*, US Environmental Protection Agency (November 1974). Report No. EPA-906/10-74-002.
35. See also *Preliminary Assessment of Suspected Carcinogens in Drinking Water*, US Environmental Protection Agency, Office of Toxic Substances (June 1975), Report No. 560-4-75-003.
36. See P. S. Ward, 'Carcinogens Complicate Chlorine Question', *Journal of the Water Pollution Control Federation*, Vol. 46 (1974), pages 2638–40.
37. 'The Isolation and Concentration of Organic Micropollutants from Drinking Water', Water Research Centre, Medmenham Laboratory. Note distributed during Open Days (4–6 May 1977).
38. *The Bacteriological Examination of Water Supplies*, Department of Health and Social Security, Government Reports on Public Health and Medical Subjects, Report No. 71 (London: HMSO, 1969).
39. S. F. B. Poynter, J. S. Slade and H. H. Jones, 'The Disinfection of Water with Special Reference to Viruses', *Water Treatment and Examination*, Vol. 22 (1973), No. 3, pages 194–208.
40. D. G. Miller, 'Research Aspects: Water', in *Symposium on Developments in Water and Sewage Treatment*, Institution of Water Engineers and Scientists (London: IWES, 1976).
41. Water Research Centre, Stevenage Laboratory, Elder Way, Stevenage, Herts. (telephone: 0438 2444).
42. Environmental Research Programme, 200 Rue de la Loi, 1040 Brussels, Belgium. Quote reference no: EUCO/MDU/73/76XII/476/76.

8 River pollution monitoring

1. *Report of a River Pollution Survey of England and Wales 1970, Volume 1*, Department of the Environment (London: HMSO, 1971). *Report of a River Pollution Survey of England and Wales 1970, Volume 2, Discharges and Forecasts of Improvements*, Department of the Environment (London: HMSO, 1972). *River Pollution Survey of England and Wales Updated 1973*, (London: HMSO, (1975). *Report of a*

River Pollution Survey of England and Wales 1973, Volume 3, (London HMSO, 1974). *River Pollution Survey of England and Wales Updated 1975* (London: HMSO 1978).
2. *Register of Research (annual) Part IV Environmental Pollution,* Department of the Environment Library, available from 2 Marsham Street, London SW1.
3. Ordnance Survey maps are available from Cook Hammond and Kell Ltd., 22–24 Caxton Street, London SW1 (telephone: 01 222 2466) and from local Ordnance Survey agents.
4. Yorkshire Water Authority, *Water Quality Inheritance on 1st April 1974.*
5. J. Cranfield and M. Bonfiel, *Waterways Atlas of the British Isles* (Pinner: Cranfield and Bonfiel Books, 1966).
6. From: J. H. N. Garland and I. C. Hart, 'Water Quality Relationships in the River System', Symposium of the Trent Research Programme organised by the Institute of Water Pollution Control, University of Nottingham (15–16 April 1971). Available from Water Research Centre as Reprint No. 621.
7. Second Annual Report of the Welsh National Water Development Authority, year ended 31 March 1976.
8. 'The Effect of the 1976 Drought on the Fisheries of Yorkshire', Paper given to the Yorkshire Water Authority Fisheries Advisory Committee (14 February 1977).
9. Source: *Water Pollution Control* (1972), page 296.
10. *Water Quality 1975–76,* Severn Trent Water Authority, page 466.
11. A. L. Downing and R. W. Edwards, 'Effluent Standards and the Assessment of the Effects of Pollution on Rivers', in 'Symposium on Effluent Standards', *Water Pollution Control* Vol. 68 (1969) No. 3, pages 283–99.
12. H. A. C. Montgomery and I. C. Hart, 'The Design of Sampling Programmes for Rivers and Effluents', *Journal of the Institute of Water Pollution Control No. 1* (1974). Available from the Water Research Centre as Reprint No. 725.
13. 'Inter-laboratory testing', Paper given to the Yorkshire Water Authority Water Quality Advisory Panel (12 February 1976).

9 *The pollution audit*

1. C. Medawar, *Social Audit Consumer Handbook* (London: Macmillan, 1978).
2. J. Catlow and C. G. Thirlwall, *Environmental Impact Analysis,* Department of the Environment, Research Report 11 (1976).
3. *Safety Officers: Sample Survey of Role and Functions,* Health and Safety Executive Discussion Document (1976).

Index

Abstract journals
 how to use, 38–40
 on toxic hazards, 39–40
 on air pollution, 49–50
 on water pollution, 57–8
 Chemical Abstracts, 23
 CIS Abstracts, 38–9
 Air Pollution Abstracts, 49
Acetaldehyde, 13
Acetic acid, 128
Acetone, 13
Acids
 particles in air, 97
 in effluents, 128
Acrolein, 46
Acute effects, 4, 20, 50, 52, 54, 167
Adhesives, 169
Aeroallergens, 46
Aerosols, 46, 97, 169
Air Pollution Abstracts, 49
Air pollution control equipment, 49, 50, 62, 73–4, 83, 108, 112
 and Alkali Inspectorate, 64–6
 and local authorities, 69–72
 see also Pollution control
Air pollution from domestic fires, 75, 78, 79, 96, 102–4, 106, 107, 110, 112
Air pollution emissions
 limits for registered works, 63–9
 limits for non-registered works, 69–71
 overseas limits, 43, 66–7
 adequacy of control, 62, 84, 162, 164
 predicting size of, 73–5, 185
 estimating impact at ground level, 62, 73–83, 100–1, 165–7

 sampling emissions, 64–5, 72, 164
 publication of sampling results, 61, 67–9, 72
Air pollution hazards,
 sources of information, 43–50
Air pollution law, 61–72, 191
Air pollution monitoring, 95–113
 National Survey, 7, 79, 95–113
 methods, 43, 49, 103–4
 moss bags, 96–7
 deposit gauges, 96, 99, 101
 continuous, 42, 101, 106, 112
 required by planning authority, 70
 location of sites, 79, 82, 95, 100–1, 107, 110, 111
 to confirm predicted impact, 81–2, 83, 84, 88, 112
 to pinpoint pollution source, 100–1, 109–10
 averaging time, 80–1, 83, 93–4, 95, 101, 105–6, 110, 111, 113
 no. of samples needed, 104–5
 adequacy, 165
 accuracy, 95, 108–9, 191
Air pollution research, 49
Air quality
 standards as alternative to emission limits, 62, 84
 objectives and criteria defined, 85
 proposed UK objectives, 88–90
 international objectives and standards, 43–4, 49, 86–7, 90–4
 criteria, 45–9
Air Quality Criteria reports, 45–6
Aldehydes, 46
Aldrin, 56
Alkali etc. Works Regulation Act 1906, 63, 184

Index

Alkali Inspectorate
  duties, 62–9
  integrated into Health and Safety Executive, 62, 63, 185
  views of Royal Commission, 63, 64, 67
  chimney height policy, 61, 63, 76, 80, 87, 187
  and nuisance, 61, 63, 65, 72
  information available from, 26, 73, 111, 184
  secrecy of, x, 61, 67, 88
  criticised, 67
Alkalis, 128, 143
Aluminium, 143
Alyn, River, 156
American Conference of Governmental Industrial Hygienists, 12, 13, 28
American Manufacturing Chemists Association, 37
Ammonia
  and air pollution, 46
  in effluents, 120, 128
  in rivers, 56, 120, 122, 130, 154–5
  drinking water limits, 142
Anglian Water Authority, 133, 141–2, 151
Antimony, 143
Arsenic
  and air pollution, 46
  in effluents, 120, 128
  and aquatic life, 4, 54, 56
  drinking water limits, 142
Asbestos
  and air pollution, 46, 47, 48
  air quality objectives, 90
Avon Medicals Ltd, xi, 70
Avon Rubber Company Ltd, x, xi

Bacteria, *see* Micro-organisms
Barium, 46, 142
Bathing, 51, 140
Bedford, 98–100, 102
Benzene, 4, 7
Benzo(α)pyrene, 73
Beryllium, 46
Best practicable means, 61, 63–6, 72, 90

Biochemical oxygen demand (BOD)
  in rivers, 122, 130, 133–4, 154–5, 156, 160
  consent conditions for, 119, 120, 126
  drinking water limit, 143
Biological aerosols, 46
Biological classification of rivers, 135
Biologic Effects of Atmospheric Pollutants, 46, 48
Birmingham City Council, 70
Bolsover, 110
Boron, 46
British Leyland, xi
British Society for Social Responsibility in Science, 25
British Standards Institute, 185
Bulgaria, 92
2-Butanone, 13, 15

Cadmium
  and air pollution, 46, 66
  in effluents, 120, 128
  in rivers, 56, 138
  drinking water limits, 142
Calcium, 143
Canada, 92
Cancer causing substances
  no safe level, 5–6, 10–11, 15
  in the air, 44, 47, 73
  in water, 54
  in drinking water, 132, 145
  in food, 182
  Avon Rubber Company policy, xi
Carbon monoxide
  TLV, 13
  and air pollution, 46, 86, 87, 90, 91, 105–6
Ceiling values, 13, 14
Central Electricity Generating Board, 81
Central Electricity Research Laboratory, 101
Central Unit on Environmental Pollution, 44, 46, 48, 182, 183, 186, 187, 189
Chemical Abstracts, 23
Chemical classification of rivers, 122, 133–4

Chemical dictionaries, 20, 169
Chemical formula, 21, 27
Chemical manufacturers
  information about toxicity of products, x, 24–5
  information about trade name products, 22, 25
  duty to test products, 24
Chemical oxygen demand (COD), xi, 119, 130
Chemline, 23, 40–1
Chimney height control, 65, 76–83, 186–7
  by local authorities, 61, 69–70, 76, 80, 87
  by Alkali Inspectorate, 61, 63, 76, 80, 87, 187
  'acceptable ground level concentrations', 77–81, 84, 87–8, 165–7
Chimney Heights Memorandum, 80–1
Chloride, 142
Chlorine
  and air pollution, 46
  in rivers, 56
  in effluents, 120, 128
  drinking water disinfectant, 51, 140, 145
Chlorine trifluoride, 14
Chlorofluorocarbons, 48
Chloroform, 142–3, 145
Chloropicrin, 14
Chromium and chromates
  and cancer, 15
  synergism with nickel, 4, 138
  and air pollution, 46, 48
  in rivers, 56
  in effluents, 120, 128
  drinking water limits, 142
Chronic effects, 4, 20, 50, 52, 54, 167
CIS Abstracts, 38–9
Citric acid, 128
Clean Air Acts 1956 and 1968, 61, 69–71, 96
Clean Air (Emission of Grit and Dust from Furnaces) Regulations 1971, 185
Clean Air Enactments (Repeals and Modifications) Regulations 1974, 184

Clean Air Law and Practice, 71
Clinical Toxicology of Commercial Products, 29, 34
Coal, SO_2 emissions from, 74–5, 96, 103
Coalite and Chemical Products Ltd., 110
Coarse fish, 135, 136, 139
COD, *see* Chemical oxygen demand
Colour, in drinking water, 141, 143
Company policies, x–xi, 163, 168
Conductivity, 143
Consent conditions
  for air pollution emissions, 90
  for discharges to water, *see* Effluent discharges
Control of Atmospheric Pollution (Appeals) Regulations 1977, 184
Control of Atmospheric Pollution (Exempted Premises) Regulations 1977, 184
Control of Atmospheric Pollution (Research and Publicity) Regulations 1977, 184
Control of Pollution Act 1974
  disclosure of information on air pollution, xi, 67–9, 70, 72
  disclosure of information on river pollution, xi, 117, 124–5, 130, 150
  'practicable' defined, 64
  effluent discharges controlled, 117, 118, 121
  public right of prosecution, 117, 121, 192
Control of Smells from the Animal Waste Processing Industry, 185
Conversion factors, 130, 175–6
Cooling water, 127, 152
Copper
  effects on fish, 56, 137–8
  in effluents, 120, 128, 157–8
  drinking water limits, 142
Costs
  taken into account in UK standards, 12, 93
  in TLVs, 13, 15–16
  in best practicable means, 64, 65
  in US standards, 91

in USSR standards, 92
of toxicity testing, 9–10
of pollution controls, 49, 64, 83, 163, 164
of cleaning polluted rivers, 133, 135
of air pollution damage, 49, 186
of disclosure of information, 68
Cupolas, 71
Cyanides
    effect on sewage treatment, 123
    in effluents, 120, 128, 129
    and drinking water, 140, 142
Cyclohexanone, 3
Czechoslovakia, 92

Dangerous Properties of Industrial Materials, 28, 34
Dark Smoke (Permitted Periods) Regulations 1958, 184
Delaney clause, 182
Deposit gauges, 96, 99, 101
DDT, 56, 128
Detergents, 56, 134, 142, 170
Development Documents for Effluent Limitations Guidelines and New Source Performance Standards, 127, 192–3
Dieldrin, 56
Directory of British Associations, 23
Disclosure of information
    on factory hazards, x, xi, 22–6, 42
    on air pollution, xi, 65–9, 70, 72
    Alkali Inspectorate secrecy, x, 61, 67, 88
    on river pollution, xi, 117, 124–5, 130, 150
    on drinking water quality, 125, 146–8
    and trade secrets, 67, 68, 70, 124
    company policies, x–xi, 163, 168
Dispersion of pollution *see* Chimney height control
Dissolved oxygen (DO)
    in rivers, 122, 130, 134, 154, 156–8
    effects on fish, 56, 119, 156
    drinking water limits, 143
Diurnal fluctuations, 154, 156–8, 160
Documentation of the Threshold Limit Values for Substances in Workroom Air, 7, 14–16, 27–8, 32

Doncaster, 104, 110
Dose-response relationship, 5, 8–9, 10, 85
Drinking water
    standards for chemical contamination, 141–6
    standards for micro-organisms, 145–6
    overseas standards, 52, 54, 141
    and cancer, 132, 145
    and river quality, 51, 122, 132, 135, 140, 144–5
    chlorination, 140, 145–6
    monitoring results published, 125, 146–8
    sources of information on hazards, 50–8
Drought, 122, 132, 134, 156
Dust, *see* Grit and dust

E. Coli, 145–6
EEC
    Environment Programme, 90
    Environmental Research Programme, 147
    Press and Information Office, 90, 140, 188
EEC Directives and proposals
    on air quality standards, 84, 85, 90, 92
    limiting sulphur in gas oil, 71, 185
    for effluent discharges, 119
    to protect drinking water, 132, 141–4, 146
    to protect drinking water sources, 122, 133
    to protect bathing waters, 140
    to protect fish, 132, 136–7, 147
    to protect other waters, 140
Effective chimney height, 77–9
Effluent discharges
    sources of information on hazards, 50–8
    predicting impact of, 128–31, 135–9, 149, 153, 166–7
    and drinking water quality, 51, 122, 132, 135, 140, 144–5
    under Control of Pollution Act, 117, 118, 121

under Rivers (Prevention of Pollution) Acts, 118, 119–20
  discharges to sewers, *see* Sewage works, Sewage treatment
  cooling water, 127, 152
  limits on volume, 119, 123
  adequacy of control, 121, 162, 164
  details of quality published, 124–5, 130, 149
Effluent monitoring
  scope of, 118, 149, 159
  location of sampling points, 150–1, 159
  adequacy of, 164
  results published, 124–7, 130
  see also River pollution, River pollution monitoring, River quality, Sewage treatment, Sewage works
Employers, duty to inform workers, 23–4, 42
Encyclopaedia of Occupational Health and Safety, 27, 30–1
Endrin, 56
Enforcement
  of workers' right to information, 24, 42
  of air pollution standards, 62, 64–6, 70, 71–2
  of river pollution standards, 118–21, 123, 130
Environment, Department of
  on air pollution committees, 68–9
  responsibility for Alkali Inspectorate, 63
  and Royal Commission on Environmental Pollution, 188
  on odours, 185
  River Pollution Survey of England and Wales, 122, 125, 133–5, 150–1
  on EEC standards, 90, 136–7
  on pollution standards, 12, 62, 84, 90
  Environmental impact analysis, 162
  Central Unit on Environmental Pollution, 44, 46, 48, 182, 183, 186, 187, 189
  and planning inquiries, 70
  map library, 188
  Register of Research, 99, 150, 186
  other publications, 48–9, 57, 80–1, 185
Environment, Secretary of State for
  and complaint against registered works, 65, 72
  and disclosure of pollution data, 68, 124
  and consent applications, 125
  Minister's views on EEC standards, 136–7

Environmental and Industrial Health Hazards: A Practical Guide, 29, 35
Environmental Health Officers, 64, 70, 71
  see also Local authorities and air pollution
Environmental impact analysis, 162
Epidemiological surveys, 7, 11, 16–17
Estuaries, 117
Ethyl chloride, 13
Ethylene, 46
Explosive hazards, ix, 55
European Inland Fisheries Advisory Commission, 55, 56, 122, 136, 147
European Standards for Drinking Water, 141–4, 146
Eutrophication, 122

Factory hazards
  contrast with air pollution hazards, 50, 87–8
  monitoring, 23–4, 41–2
Factory Inspectorate, 12, 24, 42, 163
  see also Health and Safety Executive
Fats, 128
Ferodo Ltd, 126
Fertilisers, 141, 144, 155
Finishing products, 170
First aid measures, 24, 29, 37
Fish
  indicator of river quality, 122, 134, 135, 139, 147
  killed by pollution, 147, 156, 163
  toxicity of pollutants to, 50–8, 137–9
  standards to protect, 135–7

Floods, 122
Fluorides
  and air pollution, 44, 48
  in effluents, 128
  drinking water limits, 142
Formaldehyde
  and air pollution, 46
  in effluents, 126, 128
Fume
  defined, 71
  damage to forests, 44
Furnaces, control of pollution from, 61, 69, 70–1, 74–5

Game fish, 135, 136, 139
Garnett, A., 190
Government Printing Office (US), 172
Grease, 128
Great Ouse River Authority, 150
Greater London Council, 87, 92
Grit and dust
  defined, 71
  emission limits, 61, 62, 70–1
  control equipment, 69, 73
  monitoring, 96, 99, 101
Ground level pollution, *see* Chimney height control
Guides for Short-term Exposures of the Public to Air Pollutants, 45, 47

Handbook of Environmental Control, 127, 128
Hard water, 138, 142
Hazards of Work: How to Fight Them, The, 42
Health and Safety at Work Act 1974
  duty to inform workers, x, xi, 23, 24, 42
  duty to test substances used at work, 9, 24
  control of air pollution, 61, 63–6
  prohibition notices, 65
  replaces Alkali Act, 63
  and secrecy of Alkali Inspectorate, 67
  disclosure of air pollution data, 69

Health and Safety Executive
  address, 37
  notification scheme for toxicity of new substances, 8, 182
  and TLVs, 12, 16
  monitoring of factory air, 42
  duty to inform workers, xi, 22, 24
  Alkali Inspectorate incorporated, 62, 63, 185
  Factory Inspectorate, 12, 24, 42, 163
  report on safety officers, 163
  Technical Data Notes, 37
  Guidance Notes, 12, 37
Health and Social Security, Department of, 146
Health Research Group (US), 182
High Court, 72
Hydrocarbons
  and air pollution, 46, 47
  air quality objectives for, 90, 91
  monitoring of, 97
  in effluents, 128
Hydrochloric acid, 47
Hydrogen chloride, 47
Hydrogen fluoride, 47
Hydrogen peroxide, 128
Hydrogen sulphide, 47, 142

ICI, 24
Incinerators, 71
Instab, 56–7
International Agency for Research on Cancer, 44
International Labour Office, 27, 30–1
Inversion *see* Temperature inversion
Iron, 47, 120, 142
Isomers, 21, 30
Israel, 92
Italy, 92

Lawther, P., 5
LC50 *see* Lethal concentration
Lead
  in the environment, 48
  and air pollution, 44, 47, 48
  air quality objectives, 85, 87, 90, 92
  and water pollution, 56, 120, 128
  in drinking water, 132, 141, 142, 144

in factory air, 92, 95, 96

Lethal concentration, 8–10, 137–9
Local authorities and air pollution
 responsibilities, 61, 62–3, 69–72
 and Clean Air Acts, 61, 69–71
 and Public Health Acts, 69, 71–2
 and planning controls, 69, 70, 82, 89, 91, 162
 and registered works, 65, 72
 sampling emissions, 67–9, 72
 air quality objectives, 88–90
 monitoring air quality, 96, 99, 111, 112
 as source of information, xi, 26, 61, 67–9, 72
 see also Chimney height control, by local authorities
Local authorities and water pollution, 118, 151

Magnesium, 56, 142
Manganese
 TLV, 13
 and air pollution, 47, 48
 and water pollution, 56
 drinking water limits, 142
Maps
 of air pollution monitoring sites, 97, 100, 107
 of smoke and SO_2 concentrations, 87, 99
 of river quality, 133, 150–1, 152, 159
Marine pollution, ix, 52, 117
Mercaptans, 128
Mercury
 in the environment, 48
 and air pollution, 47
 and water pollution, 56
 drinking water limits, 142
Mersey, River, 126
Metal plating industry, 128, 138
Metals
 in the air, 96
 in rivers, 152
 in sewage effluent, 123, 130
 consent conditions for, 119, 120
Meteorological Office, 79, 100

Methaemoglobinaemia, 141
Methyl ethyl ketone (2-Butanone), 13, 15
Micro-organisms
 in the air, 46
 in sewage treatment, 120, 123, 156
 in drinking water, 145–6
 in river purification, 119, 152, 154–5
Micropollutants, 132, 145
Minerals and mineral oils, 143
Mixtures of chemicals
 synergism of, 4, 138
 toxicity not investigated, 19, 85
 in trade name products, 22, 25
Molecular formula, 21
Molecular weight, 74, 176
Monitoring, see Air pollution monitoring, Drinking water, Effluent monitoring, Factory hazards, monitoring, River pollution monitoring
Moss bags, 96–7
Mutagens, 10, 54

National Environment Research Council, 186
National Institute for Occupational Safety and Health (US) *Registry of Toxic Effects of Chemical Substances*, 20, 22, 28, 33, 55
 criteria documents, 37
National Society for Clean Air Yearbook, 49, 71, 73, 99
National Survey of Air Pollution, 95–113
National Technical Information Service (US), 172
National Water Council, 121, 122, 135
NATO, Committee on the Challenges of Modern Society, 45, 46
Natural gas, 103
Nene, River, 140
New Orleans, 144–5
Nickel
 and air pollution, 47, 48
 in effluents, 120, 128

drinking water limits, 143
synergism with chromium, 4, 138
Niobium, 9
Nitrates and nitrites
　in drinking water, 132, 141, 142, 143, 144
　in rivers, 130, 154
Nitrogen
　in effluents, 120
　in rivers, 154–5
　drinking water limits, 143
Nitrogen oxides
　and air pollution, 8, 11, 46, 47, 86, 87
　air quality objectives for, 90, 91
　monitoring, 97
　sources of, 75
No-effect level, 5, 54
　see also Safe limits
Non-registered works, 69–72
　see also Local authorities and air pollution
Northampton, 140
Northumbrian Water Authority, 151
North West Water Authority, 125, 126
Norway, 80
Notes on Water Research, 55
Nuclear hazards, ix
Nuisance
　from registered works, 63, 65, 66, 112
　from non-registered works, 61, 69–70, 71–2, 90, 112
　from smoke, 86
　and air quality objectives, 88, 93
　from rivers, 122, 134

Odours
　air pollution nuisance, 47, 71–2, 185
　river pollution nuisance, 122, 134
　and drinking water, 140, 141, 143
Oil Fuel (Sulphur Content of Gas Oil) Regulations 1976, 185
Oils
　sulphur content of, 71, 74–5, 103
　in effluents, 120, 128, 152
Ordance Survey maps, 151

ORD Publications Summary, 57
Organic pollution, 119, 120, 154
Organisation for Economic Co-operation and Development, 186
Organochlorine compounds, 143, 144–5
Ozone
　and chlorofluorocarbons, 48
　air quality objectives for, 87
　monitoring of, 97
　as disinfectant, 145

Paints, 170
Particulate polycyclic organic matter, 48
Particulates, *see* Smoke
PCB, 56
Penguin Dictionary of British Natural History, 51
Permanganate value, 119, 120, 126
Pesticides
　and air pollution, 47
　and water pollution, 52, 55, 56, 141
　drinking water limit, 143
　trade directories, 170
pH
　in effluents, 119, 120, 126
　effects on fish, 53, 56, 138
　drinking water limits, 143
Phenols
　in effluents, 120, 126, 128
　effects on fish, 56
　in drinking water, 140, 142
Phosphorous, 47, 143
Photochemical oxidants
　and air pollution, 44, 46, 47, 86
　air quality objectives for, 87, 90, 91
Photochemical smog, 97
Photosynthesis, 156, 157
Planning controls and air pollution, 69, 70, 82, 89, 91, 162
Plants and air pollution, 44–8
Plastics industry, 170
Plume rise, 187
　see also Chimney height control
Poisoning by Drugs and Chemicals. An Index of Toxic Effects and Their Treatment, 30, 36
Pollens, 46

Pollution control
  aim of, 4
  adequacy of, 162, 164, 165, 167
  see also Air pollution control equipment, Sewage Treatment
Pollution Detection and Monitoring Handbook, 43, 55
Pollution law books, 190
Polycyclic aromatic hydrocarbons (PAH), 142
Potable water, 135
  see also Drinking water
Potassium, 143
Potassium permanganate, 119, 143
Power stations
  and Alkali Inspectorate, 62, 65
  impact of emissions, 79, 81-2, 101
  sulphur content of fuels, 71, 75
Practicable, 12, 16, 23, 64, 71, 93
  see also Best practicable means
Prayer, and air pollution control, 61
Preliminary Air Pollution Survey reports, 45-6
Prior approval
  for registered works, 66, 184
  for non-registered works, 69-70
Prohibition notices, 65
Proposed Scheme for the Notification of the Toxic Properties of Substances, 182
Propylene oxide, 13
Public analyst, 112
Public Health Act 1936, 69, 71-2
Public Health Act 1961, 192
Public Health (Drainage of Trade Premises) Act 1937, 192
Public Health (Recurring Nuisances) Act 1969, 185
Public interest, grounds for non-disclosure, 68, 124
Public Interest Research Centre, v, vi

Radioactive substances, 47
Rainwater, contaminants in, 147
Recycling, of effluents, 123
Registered works, 62-9
Registry of Toxic Effects of Chemical Substances, 20, 22, 28, 33, 55
Respiration, 156, 157

Ringelmann chart, 71
River authorities, 118, 121, 133
River pollution
  see also Effluent discharges, Effluent monitoring, Sewage treatment, Sewage works, Water pollution
River pollution law, 117-24, 191
River pollution monitoring
  methods, 43, 45, 158-9
  accuracy, 159-60
  adequacy, 157-8, 165
  location of sampling points, 150-1, 152-3, 159
  and short-term fluctuations, 147, 154, 158, 160
  time of sampling, 153, 156-8, 160
  results published, 124-7, 149-50, 154, 160
River Pollution Survey of England and Wales, 122, 125, 133-5, 150-1
River purification boards, 118, 173-4
River quality
  classifications, 121-2, 132, 133-5, 147
  absence of UK standards, 133
  National Water Council proposals, 121, 122, 135
  EEC standards, 119, 122, 132, 133, 136-7, 140-4, 146, 147
  EIFAC criteria, 55, 56, 122, 136, 147
  US standards, 52, 54
River quality objectives, 62, 119, 121, 135, 147, 167
  to protect drinking water, 122, 132, 141-4, 146
  to protect drinking water sources, 122, 133
  to protect bathing waters, 140
  to protect fish, 122, 132, 135-9, 147
  to protect others, 139-40
  predicting impact of discharges on, 128-30
Rivers (Prevention of Pollution) Acts 1951 and 1961, 118, 119-21, 133
Rochdale, 140
Royal Commission on Environmental Pollution
  views on Alkali Inspectorate, 63, 64, 67

Index

proposals for 'air quality guidelines', 84, 88–90, 93
Rubber industry, 171

'Safe' limits, 3–11, 16–17, 18
  for factory hazards (TLVs), 12–17, 26–43
  for air pollution, 11, 26, 43–50, 62, 76, 80–1, 83, 84–94
  for fish life, 26, 50–8, 122, 134–9, 147–8, 194
  for drinking water, 26, 50–8, 141–5, 147–8
Safety officers, 163, 168
Salmon
  and air pollution, 80
  and river pollution, 135, 136
Scheduled processes, 62–3, 183
Science Reference Library, 26, 37, 40, 41
Scorer, R.S., 109
Sea, pollution of, ix, 52, 117
Secrecy, *see* Disclosure of information
Selenium, 47, 48, 142
Severn Trent Water Authority, 125, 151, 157
Sewage treatment
  of trade effluents, 117, 121–4, 130, 154
  biological process, 120, 123, 156
  effects of chemicals on, 56–7, 123
  cost of, 133
  and drinking water, 132, 140, 145–6
  see also Effluent discharges
Sewage works
  consent limits for, 117, 121
  overloading, 123, 152, 155–6
  maps published, 151
  details of effluent quality published, 124–5
Sheffield, 105
Silica, 143
Silver, 142
Smog, 4, 79, 97
Smoke ('Particulates')
  hazards of, 5, 7, 46, 86–7
  effects on plants, 44
  synergism with SO_2, 4, 5
  air quality objectives for, 90, 91, 94

legal control of emissions, 61, 62, 69–71, 96
calculation of quantities emitted, 73, 102–4, 189
monitoring at ground level, 7, 96–113
contour maps published, 87, 99
Smoke control areas, 70, 97, 100, 107, 112
Smokeless fuel, 75, 96, 103
Smoke/SO_2 ratio, 98, 102–4
Social Audit
  objectives, v
  on Avon Rubber Company, x–xi, 70
  on Alkali Inspectorate, 67
  consumer handbook, 162
  contents of reports, 177–80
Social auditing, 161–2
Sodium, 143
Soft water, 138, 144
Solvents, 171
Starch, 128
Strontium, 56
Sugars, 128
Sulphate, 142
Sulphide and sulphite, 20, 128
Sulphur content of fuels
  figures for, 75, 102–3
  in calculating SO_2 emissions, 68, 70, 73–5
  legal limits, 71
Sulphur dioxide (SO_2)
  hazards of, 5, 6, 7, 46, 86–7
  effects on plants, 44, 48, 80, 86
  synergism with smoke, 4, 5
  air quality objectives for, 87, 90, 91
  legal control of emissions, 71, 74, 96
  calculation of quantities emitted, 68, 70, 73–5, 102–4, 189
  domestic emissions, 75, 79, 96, 102–4, 107, 110, 112
  dispersal from chimneys, 71, 76–82, 186
  monitoring at ground level, 7, 79, 81–2, 96–113
  contour maps published, 87, 99
Sulphuric acid
  hazards of, 6

manufacture, 81
monitoring in air, 80, 97
effects on sewage works, 123
Suspended solids
  in effluents, 119, 120, 126
  in rivers, 56, 134, 152, 155
Sweden
  TLVs, 13
  and UK air pollution, 44, 80, 186
Synergism, 4, 85
  chromium and nickel, 4, 138
  smoke and sulphur dioxide, 4, 5
Synonyms, 20–1, 28, 40

Tame, River, 154, 155
Taste, of drinking water, 140, 141, 143
Technical Data Notes, 37
Temperature
  of rivers, 152, 157
  of effluents, 120
  effects on fish, 56, 138
  effects on sewage treatment, 156
  drinking water limits, 143
Temperature inversion, 50, 77–9, 83, 99, 102, 106, 108, 113, 166–7
Teratogens, 9, 54
Thalidomide, 9
Thames, River, 132
Thames Water Authority, 125
Threshold Limit Values (TLVs)
  evidence for, 3, 7, 12–17, 27–8, 32, 42–3
  in Sweden and USSR, 13–14
  and air pollution standards, 87–8, 92
Tidal rivers, 117
Tilbury power station, 76
Toluene, 13
Total organic carbon, 143
Toxicity testing
  methods and limitations, 5–11, 16–17, 18–20, 167
  of air pollutants, 50
  of water pollutants, 51, 137–9
Toxline, 37, 40–1, 49, 57
Trace elements, 56
Trade Associations and Professional Bodies in the UK, 23

Trade directories, 23, 169–71
Trade effluent, *see* Effluent discharges
Trade names of chemicals, 22–3, 28, 40, 41, 169–71
Trade secrets, 67, 68, 70, 124
Trent, River, 139
Trent River Authority, 127
1,1,1-Trichloroethane, 19, 21, 24, 29–36, 38–9, 40
1,1,2-Trichloroethane, 21
Trichloroethene, 21
Trichloroethylene, 13, 21
Trout
  and air pollution, 80
  and river pollution, 51, 135, 136, 137–9
TUC, 25
TUC Centenary Institute for Occupational Health, 25, 31
Turbidity, 142–3

Uniroyal Ltd., x
United States
  air quality criteria, 45, 46, 47, 49
  air quality standards, 43, 88, 91, 92
  American Conference of Governmental Industrial Hygienists, 12, 13, 28
  American Manufacturing Chemists Association, 37
  Atomic Energy Commission, 187
  Clean Air Act, 91
  Coastguard, 55
  Council on Environmental Quality, 188
  drinking water standards, 52, 54, 141
  Department of Health, Education and Welfare, 45, 46, 86, 186
  Environmental Protection Agency, air pollution publications, 44, 45, 47, 48, 91, 92, 182, 184, 185, 189
  Environmental Protection Agency, drinking water pollution publications, 52, 141, 144–5
  Environmental Protection Agency, water pollution publications, 52, 57, 127, 139, 192–3

Index

fish protection standards, 137, 139
Food, Drug and Cosmetic Act, 182
Government Printing Office, 172
Health Research Group, 182
National Academy of Sciences, 5, 46, 48
National Institute for Occupational Safety and Health, publications, 20, 22, 28, 33, 37, 55
National Technical Information Service, 172
New Orleans drinking water, 144–5
Publications, ordering, 172
river quality standards, 52, 54
TLVs, 12
Water Resources Scientific Information Center, 55, 56
Updating information, 19, 37–41, 48–50, 57
Uranium, 10
USSR
  factory air standards, 13–14
  air quality standards, 91–2, 93
  drinking water standards, 141

Vanadium, 47, 48, 90
Vapor-phase, 48
Vehicle pollution, 86, 92, 97, 102, 103, 107
Viruses, 145–6

Wardle Fabrics Ltd, 126
Warren Spring Laboratory
  address, 189
  National Survey of Air Pollution, 95–113
  predicting pollution dispersal, 80
  and control of odours, 185
  pollution maps, 87, 97, 99, 100, 107
Waste disposal, ix, 24, 26
Water Act 1973, 141
Water authorities
  responsibilities for drinking water, 132, 141–8
  responsibilities for river pollution, 118–25
  replace river authorities,118,121,133
  pollution monitoring by, 124–7, 149–60
  to set river quality objectives, 119, 121–2, 132, 135, 147
  open to prosecution, 117, 121
  publish river maps, 150–1
  publish water quality reports, 125–6, 130, 133, 150
  information available from, xi, 26 117, 124–7, 130, 150
  addresses, 173
  see also under individual water authorities
Water conservation, 123
Water pollution, *see* Drinking water, Effluent discharges, River pollution, River quality, Sewage treatment, Sewage works
Water Pollution Abstracts, 57
Water pollution hazards
  duty of water authorities to prevent, 118
  from drinking water, 140–7
  to fish, 135–9
  sources of toxicity information, 50–8
Water pollution research, 49, 150
Water Quality Criteria, 55, 127, 140
Water Quality Criteria 1972, 52–4, 137, 140
Water Quality Criteria Data Book, 52–4, 127
Water Research Centre
  address, 55–6
  research on micropollutants, 145
  Instab, 56–7
  WRC Information, 57
  information available from, 26, 147
Waterways Atlas of the British Isles, 151
Welsh National Water Development Authority, 125, 156
Wessex Water Authority, 151
'Wholesome'
  drinking water, 132, 141
  rivers, 133
Who's Who in the Water Industry, 118
Winter/Summer ratio, 102, 110
Working Party on the Suppression of Odours from Offensive and Selected Other Trades, 185

World Health Organisation
  on air quality, 84, 86–7, 93, 94
  on drinking water quality, 132, 141–4, 146
  other publications, 44, 182
World's Air Quality Management Standards, The, 67, 92, 93

Yorkshire Water Authority
  consent conditions, 119, 120

  sampling accuracy, 159
  cost of improving rivers, 133
  and river quality, 125, 156
  river maps, 151

Zinc
  and air pollution, 47
  and water pollution, 56, 138
  in effluents, 120, 128
  and drinking water, 141, 142